Lab Manual

SIXTH EDITION
PHYSICAL SCIENCE

bju press®
Greenville, South Carolina

The writers and publisher have made every effort to ensure that the laboratory exercises in this publication are safe when conducted according to the instructions provided. We assume no responsibility for any injury or damage caused or sustained while performing activities in this book. Conventional and homeschool teachers, parents, and guardians should closely supervise students who perform the exercises in this manual. More specific safety information is contained in the PHYSICAL SCIENCE Teacher Lab Manual Sixth Edition, published by BJU Press. Therefore, it is highly recommended that the Teacher Lab Manual be used in conjunction with this manual.

NOTE: The fact that materials produced by other publishers may be referred to in this volume does not constitute an endorsement of the content or theological position of materials produced by such publishers. Any references and ancillary materials are listed as an aid to the student or the teacher and in an attempt to maintain the accepted academic standards of the publishing industry.

PHYSICAL SCIENCE Student Lab Manual
Sixth Edition

Coordinating Writer
David M. Quigley, MEd

Writers
Christopher D. Coyle
Jeff S. Foster, MS
Elwood Groves, MA

Biblical Worldview
Tyler Trometer, MDiv

Academic Oversight
Jeff Heath, EdD
Rachel Santopietro, MEd

Editor
Rick Vasso, MDiv

Cover, Design, and Interior Concept Design
Sarah Lompe

Page Layout
Carrie Walker

Illustrator
Sarah Lompe

Permissions
Lily Kielmeyer
Hannah Labadorf
Rita Mitchell
Ashleigh Schieber
Elizabeth Walker

Project Coordinator
Chris Daniels

Photo credits appear on pages 262–63.

All trademarks are the registered and unregistered marks of their respective owners. BJU Press is in no way affiliated with these companies. No rights are granted by BJU Press to use such marks, whether by implication, estoppel, or otherwise.

The cover photos show a close-up of paint in water.

© 2020 BJU Press
Greenville, South Carolina 29609
Fifth Edition © 2014 BJU Press
Originally published as Student Activities in Basic Science © 1983 BJU Press

Printed in the United States of America
All rights reserved

ISBN 978-1-62856-506-5

15 14 13 12 11 10 9 8 7 6 5

CONTENTS

Welcome to the Laboratory! — vi

UNIT 1 THE STRUCTURE OF MATTER

CHAPTER 1   MODELING OUR ORDERLY WORLD
- **1A** **Based on a True Story** Thinking Safe in the Laboratory — 1
- **1B** **How Do We Measure Up?** Practicing Measuring — 9
- **1C** **Visual Data** Graphing Mass and Volume — 15

CHAPTER 2   MATTER
- **2A** **Has Mass, Occupies Space** Modeling Matter — 21
- **2B** **Something Old, Something New?** Detecting Physical and Chemical Changes — 27

CHAPTER 3   THE ATOM
- **3A** **The Fiery Trial** Using Flame Tests — 33
- **3B** **Big Time** Inquiring into Scale Models of Atoms — 37

CHAPTER 4   THE PERIODIC TABLE
- **4A** **Bricks and Feathers** Exploring Element Density in a Period — 41
- **4B** **How Wide Is an Atom?** Exploring Atomic Radii Trends — 45

CHAPTER 5   BONDING AND COMPOUNDS
- **5A** **The Solution to a Problem** Solubility and Chemical Bonds — 51
- **5B** **Electric Lines** Conductivity and Chemical Bonds — 55

CHAPTER 6   THE CHEMISTRY OF LIFE
- **6A** **Sticky Business** Inquiring into Glues — 59
- **6B** **Milking Chemistry** Proteins in Food — 63

UNIT 2 CHANGES IN MATTER

CHAPTER 7   CHEMICAL REACTIONS
- **7A** **Science Fair Revisited** Types of Chemical Reactions — 67
- **7B** **It's in the Bag** Inquiring into Chemical Reactions — 73

CHAPTER 8   NUCLEAR CHANGES
- **8A** **Flipping Out** Modeling Radioactive Decay — 77
- **8B** **Radioactive!** Exploring Radiation Dose — 85

CHAPTER 9   SOLUTIONS
- **9A** **All Mixed Up** Inquiring into Separating Mixtures — 89
- **9B** **That's Cold!** Investigating Freezing Point Depression — 93

CHAPTER 10   ACIDS, BASES, AND SALTS
- **10A** **pH pHun** Determining pH — 97
- **10B** **Feeling the Burn** Comparing the Concentrations of Basic Solutions — 103

UNIT 3 MATTER IN MOTION

CHAPTER 11 KINEMATICS
11A **Way to Go** Inquiring into Distance and Displacement — 107
11B **Slow and Steady** Investigating Uniform Motion — 111
11C **The Gravity of the Situation** Investigating Free Fall — 117

CHAPTER 12 DYNAMICS
12A **Lab Heard Round the World** Investigating the Second Law of Motion — 121
12B **Rough Going** Investigating the Properties of Friction — 125

CHAPTER 13 WORK AND MACHINES
13A **A Clear Advantage** Investigating Pulleys — 129
13B **Ramping Up** Experimenting with Inclined Planes — 133

CHAPTER 14 ENERGY
14A **Hold Your Horses** Investigating Work, Energy, and Power — 139
14B **Time to Climb** Designing a Better Motor — 145

CHAPTER 15 THERMODYNAMICS
15A **Metal Mystery** Inquiring into Specific Heat — 149
15B **Around the Curve** Investigating the Heating Curve of Water — 155

CHAPTER 16 FLUIDS
16A **High Pressure Job** Investigating Fluid Mass and Pressure — 161
16B **Load, Load, Load Your Boat** Designing a Paper Boat — 165

UNIT 4 WAVES AND ENERGY

CHAPTER 17 PERIODIC MOTION AND WAVES
17A **Tick Tock** Investigating Pendulums — 169
17B **Storm Surge!** Creating Coastal Defenses — 175

CHAPTER 18 SOUND
18A **Sounding Off** Investigating the Properties of Sound — 179
18B **Sound Advice** Designing a Sound-Dampening Surface — 185

CHAPTER 19 ELECTRICITY
19A **Go with the Flow** Investigating Ohm's Law — 189
19B **Series-ously?** Investigating Series Circuits — 195
19C **The Path Less Traveled** Investigating Parallel Circuits — 201

CHAPTER 20 MAGNETISM
20A **Lines of Force** Exploring Magnetic Fields — 207
20B **Mighty Magnets** Inquiring into Electromagnets — 211

CONTENTS

CHAPTER 21 ELECTROMAGNETIC ENERGY
21A Light Limit Investigating Changes in Light over Distance — 213
21B Driven to Diffraction Investigating the Bending of Light — 221

CHAPTER 22 LIGHT AND OPTICS
22A Upon Reflection Investigating Mirrors and Virtual Images — 227
22B Through the Lens of the Beholder Exploring Lenses — 231

APPENDIX
- A Laboratory and First-Aid Rules — 237
- B Safety Data Sheets (SDS) — 240
- C Laboratory Equipment — 244
- D Laboratory Techniques — 246
- E STEM Design Process — 250
- F Investigating with Labdisc — 251

PERIODIC TABLE OF THE ELEMENTS — 266

Welcome to the Laboratory!

Nothing is quite as exciting as a well-stocked laboratory. Don't believe it? Well, hopefully by the end of this course you will agree. A laboratory is anything but boring. Laboratories are places where exciting discoveries are made, old ideas are challenged, and new ideas are formed. But a laboratory can't do any of those exciting things on its own. It needs a scientist. That's you!

THINKING LIKE A STUDENT SCIENTIST

You're about to begin a new school year. And one of the most important things that you can do this year is to learn to see yourself not as a science *student* but as a student *scientist*. Believe it or not, there's a big difference between the two. A science student is just a student who happens to be studying science. A student scientist, on the other hand, is a scientist in training. Even if you don't see a science career in your future, for this year, at least, you're a scientist!

Learn to think like a scientist. Your textbook discusses in detail the way that scientists think and work. Try to think along those lines as you collect data, build models, and test your ideas. Know how to construct a hypothesis. Learn how to collect accurate and precise data. Refine your laboratory techniques as you use scientific instruments to gather data. Remember, a scientist can't expect to get good results if he's careless with his instruments.

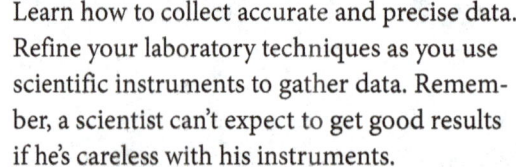

Above all, learn to see science through the lens of a Christian worldview. Everyone has a worldview. It is the lens through which you view *everything*. Christians gain their worldview from the Bible. This worldview doesn't just affect what they believe about God, Jesus, and salvation. It also affects how they view science, history, and even the way that they interact with other people.

As a student scientist, keep your worldview in mind. When you do an experiment, think about how it relates to a biblical worldview. And you're not just a student scientist. You're a *Christian* student scientist. Christians should see science as an important gift of God, given to humans so that they can obey the Creation Mandate (Gen. 1:28) and the two great commandments (Matt. 22:38–40). When Christians do science, they bring glory to God and help their fellow humans to live in this fallen world. Therefore, some questions will specifically ask you to use your worldview to apply the information from your lab activity to a real-world problem. You will be challenged to consciously think like a Christian as you're doing science.

Christians shouldn't see science as a pathway to truth. Rather, they should look to the Bible, the one source of *absolute truth*, to establish the guidelines and proper role of science in human life. Scientists working from a Christian worldview should evaluate existing scientific models in light of Scripture and challenge or revise them when they conflict. And as they create models of their own, they must test them for agreement with the revealed truths in the Bible. When we act in this way, we glorify God by using science as He intended.

KEYS TO LABORATORY SUCCESS

Lab activities can be both interesting and fun. But your laboratory experience largely depends on you. If you take these activities seriously, keep your brain engaged, and try to learn as much as possible, you'll come away from the laboratory feeling that your time has been well spent. On the other hand, if you treat lab activities carelessly, you probably won't learn much, nor will you have a genuine sense of accomplishment. The following guidelines will help you get the most from lab activities. And they'll also help you experience the true pleasure of a brain at work.

1. Read through the entire lab procedure before class. You'll know what to expect, and you'll be far less likely to make mistakes or forget a crucial step.

2. Review the parts of your textbook that apply to the lab activity. Your teacher may give you specific review guidelines.

3. Come to class well prepared. Bring your textbook, calculator, paper, and pencil. You won't necessarily need all these for every lab activity, but you'll regret it if they're sitting in your locker on a day when you *do* need them!

4. Begin the lab activity only when instructed to do so by your teacher. Read each lab procedure step carefully so that you don't make foolish mistakes. Procedure steps are identified by lettered bullets (A, B, C, etc.). If you're ever in doubt about what to do, *ask*, don't guess.

5. Lab activity questions are mixed in with the procedure steps. They're identified by numbers (1., 2., 3., etc.).

6. Many lab activities require you to record data in tables. Some also include graphing areas. Typically—but not always—data tables and graphing areas are located at the end of the activity.

7. Record your results from measurements and calculations carefully and accurately.

8. Most lab activity questions ask you to explain your answer. Don't view this requirement as pointless busywork. Explaining something makes you think more deeply about it. It also helps you connect principles from the textbook and class discussion to what you're doing in the laboratory.

9. Keep your laboratory area tidy. At the end of class, put away your equipment and dispose of trash according to your teacher's instructions.

10. While socializing may seem like a lot more fun than the lab activity, leave that behavior outside the laboratory door. You'll be more focused, get better results, and make far fewer mistakes.

Welcome to the Laboratory!

STAYING SAFE

Nothing ruins a day quite so much as getting hurt in the laboratory. While the lab activities in this manual have been carefully designed to be safe, no lab activity is perfectly safe. For this reason, you should maintain constant vigilance when you're working through one. You should always use protective equipment when it's specified.

Rather than cover laboratory safety in this introduction, we've devoted an entire lab to the topic (Lab 1A—*Based on a True Story*). This lab activity isn't just a list of do's and don'ts. It takes a novel approach to safety, one that we hope will make laboratory safety more enjoyable and more effective. Be sure to give Lab 1A your full attention so that your class can have an accident-free year!

One of the most important safety precautions you can take is to wear eye protection. Yes, goggles can be uncomfortable, but a little discomfort is worth avoiding losing an eye. Always wear your goggles anytime you are working with glassware, chemicals, or projectiles. Lab activities that require eye protection or other safety precautions will have safety icons that pertain to that activity immediately below the equipment and materials list. The icons that you will see are listed below.

 Body Protection—Chemicals or other materials could damage your skin or clothing. You should wear a laboratory apron, chemical resistant gloves, or both.

 Corrosive Substance—Acids and bases are corrosive substances that can cause damage to skin and eyes. Wear proper protective equipment and follow safe-handling procedures.

 Electricity—Electrical equipment in use. Electric shock potential exists.

 Extreme Temperature—Extremely hot or cold temperatures may cause skin damage. Use proper tools to handle laboratory equipment.

 Eye Protection—There is a possible danger to the eyes from chemicals or other materials. Wear safety goggles.

 Fire Hazard—Bunsen burners in use or flammable material in use. Use caution to avoid fire hazard.

 Sharp Object—Be careful when using equipment to avoid cuts from sharp instruments or broken glassware.

 Toxic Material—Be careful with chemicals that have been identified as toxic to humans.

Since some lab activities have specific steps with unique safety challenges, special safety notes are occasionally placed within the activity in ***bold italic font*** to emphasize their importance. For additional information on laboratory safety, see Appendix A.

LAB EQUIPMENT AND TECHNIQUES

Science involves both equipment and techniques that you may not be familiar with. To help, we have included appendixes for these. Appendix C is a visual glossary of laboratory equipment, and Appendix D explains many of the techniques that you will need for the lab activities in this manual.

INQUIRY AND STEM LAB ACTIVITIES

As you work through this lab activity book, you will find inquiry or STEM lab activities. These activities are part of being a student scientist.

Inquiry lab activities give you opportunity to create your own activity about a given subject. You will ask questions, form hypotheses, design investigations, collect and analyze data, draw conclusions, communicate results, and often develop additional questions. These are tasks that scientists do every day.

STEM lab activities give you the ability to solve real-world problems using the disciplines of science, technology, engineering, and mathematics. They will often involve collaboration with other students. An important part of STEM lab activities is the STEM design process, shown in Appendix E.

LABORATORY TECHNOLOGY

Several of the lab activities in this manual feature GlobiSens Labdisc® probeware. This technology replaces dozens of traditional laboratory instruments with a portable electronic device equipped with a wide variety of sensors. With a Labdisc, you can collect your own data and then create models from it immediately. The Labdisc works perfectly in the laboratory and also out in the field.

Since many of the lab activities in this manual use the Labdisc Gensci model, we've included a complete tutorial for using this valuable tool (see Appendix F). If your laboratory is equipped with Labdisc units, please read the tutorial right away. This tool will enable you to see scientific principles come to life in new and fascinating ways.

LET THE YEAR BEGIN

One of the pleasures that we get from writing this book is knowing that for many of you the laboratory will become a special place too. We invite you to come into the laboratory part of this course expecting to learn new and fascinating things. Look forward to the challenge of getting your hands on science. Your visit to the laboratory could be a life-changing moment for you. Welcome!

Welcome to the Laboratory!

CHAPTER 1: MODELING OUR ORDERLY WORLD

LAB 1A

BASED ON A TRUE STORY

Thinking Safe in the Laboratory

Fred Johnston felt excited as he walked into the high-school auditorium packed with eager students. Just two days ago he had graduated from college with a degree in chemistry. In a few weeks he would start a new job. But before he left college for good, he'd agreed to travel with a college team that visited high schools to get students excited about science.

Fred loved giving chemistry demonstrations. The highlight of his presentation was an "alcohol cannon," a gadget that shot toy footballs into the audience. A small explosion of ethyl alcohol vapor powered the cannon and, as an added bonus, shot a fountain of blue flames from the muzzle.

But Fred's demonstration fell sadly flat. The disappointed students left the auditorium, and Fred went to lunch unhappy. Initially he thought that the high humidity was the culprit, but as he ate lunch he began to think that the igniter might not be working.

Since he planned to do a second demonstration later that day, Fred went back to the empty auditorium to check the cannon. An hour had passed, so Fred assumed that the alcohol vapor had aired out of the cannon. Squinting down the muzzle, Fred clicked the igniter to see whether it was sparking properly. With a whoosh, blue flames erupted out of the muzzle straight into Fred's unprotected face. Shortly afterward, Fred lay in a hospital emergency room with a burned face, no eyebrows, and no eyelashes. Medical staff bandaged his eyes shut until a doctor could evaluate the damage.

QUESTIONS

How can we safely investigate in the laboratory?

- What does it mean to think safe?
- Why are there so many laboratory safety rules?
- Where is the safety equipment in our laboratory?
- Why should we read the safety data sheets (SDS)?

Equipment

safety data sheets (SDS)
(Appendix B)

PROCEDURE

Thinking Safe

Fred's accident didn't have to happen. It came about because Fred wasn't *thinking safe*. Although Fred had learned plenty of safety rules during his four years as a college chemistry student, he had never learned to think safe. Thinking safe means looking around you and trying to imagine how an accident *might* happen. In other words, you identify unsafe situations and fix them *before* they become accidents. In Fred's case, not only did he violate several safety rules—he also failed to think ahead and realize that the cannon might still be capable of harming him.

Laboratory safety isn't just about rules and regulations. While you will learn some rules during this lab activity, what we really want you to do is develop a habit of thinking safe. So even before we look at specific laboratory safety rules, let's see how attentive you are to safety.

A Take out a sheet of notepaper and write your name at the top. Add two headings: *Picture 1* and *Picture 2*.

PICTURE 1

B Look at the pictures on pages 2 and 3. Imagine that they represent your science laboratory with the lab activity having already begun. Survey the room shown in each picture. Some activities are safe and some are not. Don't just look for obvious unsafe situations. Look for accidents that *could* happen.

C Under the appropriate heading, list everything that you discover in the pictures. Explain *why* the activity or situation could be a risky one.

Science teachers and education researchers have studied high-school science laboratories and discovered something rather disturbing. More laboratory accidents occur in ninth-grade science classes than in any other grade level.

D Add to your paper the heading *Laboratory Accidents*.

E In a few sentences, discuss possible reasons for the accident statistic mentioned above. Think about the things that you have observed in your classmates' classroom behavior.

F List what you think are the top three accidents that happen in high-school science laboratories.

PICTURE 2

Based on a True Story

Oh No! More Rules!

Let's be honest—laboratory safety rules are pretty boring. Very few students begin their school year eagerly anticipating the safety discussion. And let's be even more frank—very few teachers enjoy talking about safety rules. So if everyone dislikes them, why must we keep discussing them?

The answer is simple. Safety rules are the result of *accumulated wisdom*, a fancy term meaning simply that they came about through experience and lots of accidents. Over the years, scientists, teachers, and science students have learned the hard way what *not* to do in the laboratory. Rules help you avoid repeating others' mistakes. But let's do more than memorize a list of rules. Instead, let's ask ourselves *why* these rules are valuable.

G Add another heading to your paper: *Reasons Behind the Rules*.

H Read each rule listed below. Think about each one for a moment.

I In a sentence or two, explain the reason behind each rule. If appropriate, also explain how following the rule could prevent an accident from happening.

BASIC LABORATORY RULES

1. Always keep walkways clear.
2. Don't lean on, hang over, or sit on the laboratory tabletops.
3. Know the location of all laboratory safety equipment.
4. Always wear safety goggles when working with chemicals, glassware, projectiles, or other eye hazards. Wear protective clothing and gloves when working with corrosive or staining chemicals.
5. Tie back long hair and avoid wearing loose clothing or ties.
6. Work at your own laboratory station. Do not wander around or socialize with other lab groups.
7. Never perform an unauthorized experiment or change any assigned experiment without the teacher's permission. Follow all experiment directions carefully and exactly.
8. Avoid playful, distracting, or boisterous behavior. Do not use personal technology unless it's required for the lab activity.
9. Never taste any chemical or drink out of laboratory glassware.
10. Never eat, drink, or chew gum in the laboratory.
11. To smell a substance, gently fan its vapor toward you.
12. Never lift solutions or glassware above eye level.
13. Never leave a flame or heater unattended. Keep combustible materials away from flames and heat sources.
14. Use appropriate hand protection when working with hot glassware or heat sources. *Remember*: hot glass doesn't look hot!
15. When heating a substance in a test tube, point the open end away from yourself and others. ***Never heat a closed or stoppered container!***

Laboratory eyewash stations and emergency showers can rinse off chemicals quickly.

16. When diluting acid solutions, always add the acid to water slowly. *Never add water to an acid!*
17. Do not return unused chemicals to a bottle. Dispose of them properly.
18. Notify the teacher of any injuries, spills, or breakages.
19. Always clean up your workspace prior to leaving the laboratory. Always wash your hands after cleaning your workspace.
20. Place solid trash in the designated trash container. Pour waste liquids into designated liquid waste containers.

LABORATORY SAFETY EQUIPMENT

Most school laboratories will have basic safety equipment designed to be used in an emergency. Find the following equipment in your laboratory and complete the table below. Your teacher will demonstrate how to use certain pieces of equipment.

EQUIPMENT	PURPOSE	LOCATION
Eyewash Station	to remove chemicals that may splash into the eyes	
First Aid Kit	to provide basic medical supplies for treating common injuries	
Fire Blanket	to extinguish clothing fires and fires on the laboratory bench	
Fire Extinguisher	to extinguish fires in the laboratory (Different types of extinguishers are used for different kinds of fires.)	
Shower	to remove chemicals that may splash onto clothing; to extinguish clothing fires	

HANDLING COMMON ACCIDENTS

Despite our best efforts, accidents still happen. When they do, it's crucial not to panic. Calm thinking and thoughtful actions can keep accidents from becoming worse. *Remember, all accidents or injuries must be reported to your teacher.* While your teacher will probably handle most serious situations, you should understand how common laboratory injuries are treated. For more information on laboratory first aid, refer to Appendix A.

INJURY/ACCIDENT	RESPONSE
Chemical in the Eyes	Flush with water for at least 15 minutes. If you have contact lenses, remove them after you have rinsed your eyes for several minutes. Seek medical attention.
Minor Heat Burn	Cool the area with cool water. If minor blistering occurs, apply a sterile bandage.
Chemical Burn	Flush the area with water or an approved neutralizer. Remove contaminated clothing. Seek medical attention for significant burns or allergic reactions.
Cut	Wash the area with soap and water and cover with a sterile bandage. If there is significant bleeding, apply pressure with a sterile bandage and seek medical attention.
Inhalation of Dust or Vapor	Move to fresh air. Seek medical attention if irritation or breathing difficulty occurs.
Swallowing a Chemical	Identify the chemical involved. Seek medical attention or contact a poison control center, or both.
Chemical Spill	Hazardous or corrosive chemical spills should be handled by your teacher.
Fire	Smother a small fire with a fire blanket or container (such as a beaker). For a larger fire, evacuate the area and use a fire extinguisher only if you can do so safely. When in doubt, call the fire department. Smother a clothing fire with a fire blanket or drop and roll to extinguish the flames.
Broken Glass	Do not handle broken glass with your bare hands—broken glass is extremely sharp! Sweep up fragments in a dustpan and discard in an authorized trash container.

KNOW YOUR CHEMICALS

Being informed is the best way to be safe with chemicals. Even household chemicals can be harmful under certain conditions. The best place to learn about a chemical is its label. Household chemicals provide detailed hazard and safety information right on the label. Read it carefully for each new chemical that you encounter.

When it comes to laboratory-grade chemicals, there are two sources of information—the product label and the *Safety Data Sheet* (SDS). Both of these have formats that comply with the *Globally Harmonized System* (GHS). The GHS is an international standard for providing information about chemicals, which the United States adopted in 2012. All GHS labels have six required parts. The image on page 7 highlights each of those six parts. The other source of information on laboratory-grade chemicals is a *Safety Data Sheet* (SDS). Laboratories are required to have a copy of an SDS for any chemical used in the laboratory.

J Add to your paper the heading *Safety Data Sheet*. Look at each SDS provided in Appendix B. Identify the chemicals from Section 1 of each SDS and add to your paper.

K What information is provided in Section 2 of the SDS?

L Where would you find first aid information on an SDS?

M Which chemical is an odorless white powder, a crystal, or flakes?

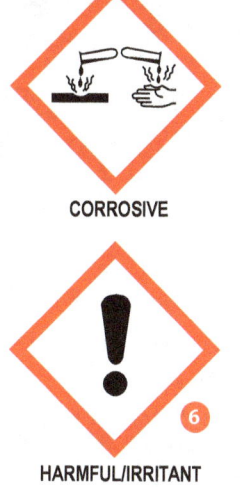

① product information
② signal word.
③ hazard statements.
④ precautionary statements.
⑤ supplier information.
⑥ pictograms

HAPPILY EVER AFTER?

As Fred Johnston lay in the emergency room, he had plenty of time to reflect. If he had just taken the time to think safe, he wouldn't be in the hospital. He'd be in the high-school auditorium doing his second demonstration.

Happily, Fred's story did have a happy ending. When the doctor examined Fred's eyes, he found them undamaged by the fireball. Other than a few minor facial burns, Fred was fine. A few hours later, the hospital released him and he returned home to recuperate.

But many laboratory accidents don't have happy endings. Some students carry long-lasting scars as a reminder of their failure to think safe. A few have to go through life without their eyesight or a limb. And some have even paid with their lives.

As you begin this school year, we hope that you'll find it a rewarding time as you explore the wonders of physical science. But we also want it to be a safe time. So keep thinking safe!

N Answer the following question on your paper: How does the Creation Mandate imply that we should always think safe?

CHAPTER 1: MODELING OUR ORDERLY WORLD

LAB 1B

HOW DO WE MEASURE UP?

Practicing Measuring

Scientists always strive to be both precise and accurate in their measurements. Precision is the degree of exactness of a measurement and is primarily dictated by the instrument used. You always measure to the next lower decimal place than the smallest marked whole decimal place on the instrument. So if your instrument is marked to the thousandth of a meter, you can measure to the ten-thousandth of a meter.

Accuracy is closeness to the accepted or expected value and depends on the scientist's technique. Often scientists describe their accuracy using percent error. But how much error can we have and still consider our measurement accurate? Some scientists may be happy with 70% accuracy. Others may think that 85%, 90%, or higher is the desired goal. In reality, it depends on what is being measured. For example, if the US Postal Service were 99% accurate in delivering mail, 1.5 *billion* pieces of mail every year—4.1 million pieces every day—would be lost!

QUESTIONS

How can we get the most accurate and precise measurements?

- What information can I find on a measuring instrument?
- Why are there different instruments to measure the same thing?
- How do beakers and graduated cylinders compare?

Equipment

laboratory balance
metric ruler
graduated cylinder, 10 mL
graduated cylinder, 100 mL
beaker, 100 mL
ring stand
ring stand clamp
metal rod
spring
mass
stopwatch
computer paper
tennis ball
pennies (5)
goggles

PROCEDURE

As you collect data, use the appropriate rows of Tables 1 and 2 (p. 14) to record your data.

A Look at your ruler. What units does the ruler measure? What do the largest markings represent? What do the smallest markings represent? Record your answers in Table 1.

1. What decimal place can you measure to with this ruler? Explain.
   _____
   _____
   _____

B Measure the length and width of the sheet of paper. Record your measurements in Table 1.

2. Is it possible to measure the thickness of the paper? If yes, how would you do that?
   _____
   _____
   _____

3. Is it possible to find an average thickness of paper? If yes, how would you do that?
   _____
   _____
   _____

C Measure the circumference and diameter of the tennis ball. Record your measurements in Table 1.

4. How does measuring the tennis ball compare with measuring the paper? Would a different instrument (not a ruler) help? What would you prefer to use? Why would it make a difference?
   _____
   _____
   _____
   _____

D Measure the diameter and height of a stack of five coins. Record your measurements in Table 1.

E Measure the mass of the five coins. Record the value in Table 1.

F Look at the graduated cylinders. What units do the graduations measure? What do the largest markings represent? What do the smallest markings represent? Record your answers in Table 1.

5. What decimal place can you measure to with each graduated cylinder?
   _____
   _____
   _____

G Add water to each of the graduated cylinders. Measure the volume of water in each of the graduated cylinders. Record the volumes in Table 1.

6. What do you notice about the top surface of the liquid in the cylinders?
   _____
   _____

This curved upper surface is caused by the attraction of the liquid molecules to the molecules of the cylinder and is called the *meniscus*. To use a graduated cylinder properly, you read the measurement at the lowest point on the meniscus. If you need to repeat the measurements from Step G, do so now. The meniscus for some liquids curves down at the edges. In this case you would read the measurement at the highest point on the meniscus.

H Add 80 mL of water to the beaker; be as exact as you can. Now pour the water into the 100 mL graduated cylinder and measure the volume of water. Record the value in Table 1.

7. Were the volumes the same? If not, how do you explain the difference?
   _____
   _____
   _____

I Push the mass on the spring upward a small amount. Release the mass and time how long it takes for one complete cycle (the mass moves from the top, where you released it, down to the bottom and back up to the top again). Then repeat, but time how long it takes for 5 complete cycles (top–bottom–top). What is the average time for one cycle in the second trial? Record the values in Table 1.

8. How does the average time compare with the time for one cycle?
   _____

9. Which do you think is more accurate: timing one cycle or timing five cycles and calculating the average? Explain.
   _____
   _____
   _____
   _____

How Do We Measure Up?   11

ANALYSIS

10. Compare your data with one other student. Were your measurements all the same? Why or why not? Which of you is correct?
   _____
   _____
   _____

J Calculate the area of the paper using the formula for the area of a rectangle. Record your answer in Table 2.

K Calculate the volume of the tennis ball using the formula for the volume of a sphere. Record your answer in Table 2.

L Calculate the volume of the coins using the formula for the volume of a cylinder. Record your answer in Table 2.

M Calculate the density of the coins. Record your answer in Table 2.

N The accepted value for the density of a penny is 6.78 g/mL. Use the percent error formula below to calculate your percent error. Record your answer in Table 2.

$$\%_{error} = \left(\frac{measured - accepted}{accepted}\right) 100\%$$

GOING FURTHER

11. If the density of copper is 8.96 g/mL, suggest a hypothesis for why the density of a penny is so much less.

   _____
   _____
   _____

12. Were you accurate in your measurements? Explain.

   _____
   _____
   _____

13. Were you precise in your measurements? Explain.

   _____
   _____
   _____

14. Compare accuracy and precision.

   _____
   _____
   _____
   _____
   _____

Table 1 DATA

	UNITS	LARGEST MARKS	SMALLEST MARKS
Ruler			
	LENGTH (cm)	LENGTH (cm)	WIDTH (cm)
Paper			
	CIRCUMFERENCE (cm)	CIRCUMFERENCE (cm)	DIAMETER (cm)
Tennis Ball			
	DIAMETER (cm)	HEIGHT (cm)	MASS (g)
Coins			
	UNITS	LARGEST MARKS	SMALLEST MARKS
Large Cylinder			
Small Cylinder			
	VOLUME (mL)		
Large Cylinder			
Small Cylinder			
	VOLUME (mL)		
Beaker	80		
Graduated Cylinder			
	TIME (s)	AVERAGE TIME (s)	
1 cycle			
5 cycles			

Table 2 CALCULATIONS

	AREA (cm²)		
Paper			
	VOLUME (cm³)		
Tennis Ball			
	VOLUME (cm³)	DENSITY (g/cm³)	% ERROR
Pennies			

14 Lab 1B

CHAPTER 1: MODELING OUR ORDERLY WORLD

LAB 1C

VISUAL DATA

Graphing Mass and Volume

We see graphs of many types all the time, and they depict all sorts of information. We have all seen bar graphs, line graphs, pie graphs and pictographs. But there are also population pyramids, spider charts, area graphs, and trellis graphs. These types can be combined to make new and different types of graphs. The key is to present data in a visual manner that clearly communicates its significance.

Understand that graphs are a type of model of the physical system that is being studied. For example, pediatricians use growth charts to monitor the growth of a child. At each visit, the doctor records the child's height and weight. The doctor then compares her patient's data with the expected growth model to assess how well the child is growing. The model is a prediction of how the child should grow before the next visit. In cases where the child doesn't grow as predicted, the doctor will look for a reason for the change in growth. Sometimes the unexpected growth is due to natural variation in how kids grow, but at other times it may indicate a serious medical condition. By collecting and graphing data and comparing it with the predictions of the model, doctors can detect problems early and intervene.

QUESTIONS

How much different information can I find on a graph?

- How do I create a graph?
- What information can I get from a graph?
- How are values on the graph related to the system I am studying?

Equipment

laboratory balance
graduated cylinder, 100 mL
beaker, 250 mL
disposable pipette
salt water solution (15%), 100 mL
goggles

PROCEDURE

Matter is anything that has mass and takes up space. As can be seen on a growth curve, we expect children's mass and weight to increase as they grow taller and take up more space.

1. What is mass?

   _____

2. You have a sample of iron that occupies 50 cm³ of volume. Someone adds more iron so that the mass is doubled. How much space does the iron occupy now? Explain.

   _____

   _____

   _____

A Add approximately 20 mL of the saltwater solution to the graduated cylinder. Measure the exact volume of the solution and record the measurement in the *Trial 1* row of Table 1 (p. 19).

B Measure the mass of the graduated cylinder and solution. Record the measurement in the *Trial 1* row of Table 1.

C Add an additional 20 mL of the solution to the graduated cylinder. Measure the exact volume of the solution and record the measurement in the *Trial 2* row of Table 1.

3. Step C approximately doubled the volume of solution. Using the mass measured in Step B, what would you conclude about the mass of the graduated cylinder now?

   _____

D Mass the graduated cylinder and solution. Record the measurement in the *Trial 2* row of Table 1.

4. Why didn't the mass double even though you doubled the volume of the solution?

   _____

   _____

E Add an additional 15 mL of solution to the graduated cylinder. Measure the exact volume of solution and record the measurement in the *Trial 3* row of Table 1.

F Mass the graduated cylinder and solution. Record the measurement in the *Trial 3* row of Table 1.

G Add an additional 15 mL of solution to the graduated cylinder. Measure the exact volume of solution and record the measurement in the *Trial 4* row of Table 1.

H Mass the graduated cylinder and solution. Record the measurement in the *Trial 4* row of Table 1.

I Graph your volume data (*x*-axis) and mass data (*y*-axis) in the graphing area (p. 20). Include a curve of best fit through the data points. Make sure that you include a graph title and axes labels (including units). For help with graphing, refer to Appendix B of your textbook.

ANALYSIS

Scientists use their data and graphs to learn more about the system they are studying. Scientists will often determine equations, slopes, and *x*- and *y*-intercepts from their graphs. These mathematical quantities often relate to physical characteristics of the system they are studying.

5. Using the slope formula, calculate the slope of the curve of best fit. Circle your final answer.

$$m = \frac{\Delta y}{\Delta x} = \frac{y_2 - y_1}{x_2 - x_1}$$

6. The changes in mass and volume were due only to adding more solution. So then what does the slope that you calculated represent?

_____

Scientists often use their data and graphs to predict values that were not part of an original data set. Mathematical models, such as your curve of best fit, are evaluated on the basis of how good they can predict other values. Sometimes scientists predict values within the range of their data values using *interpolation*. For example, a doctor with data on one- and two-year-old children could estimate, or interpolate, information about a child that is a year-and-a-half old. Let's see how this works on your graph.

J Using your graph, predict the volume of solution that would make a graduated cylinder and solution mass of 140.00 g. Record your prediction in Table 2 (p. 19).

K Empty the graduated cylinder, place it on the balance, and add solution until you reach a mass of 140.00 g. Use the pipette if needed to get the mass exact. Measure and record the actual volume in Table 2.

L Calculate the percent error between your predicted volume (Step J) and the actual volume (Step K). Record the percent error in Table 2.

At other times scientists predict values outside the range of their data values using *extrapolation*. Typically growth curves include data through the age of 18. To use this data to estimate the height and weight of a 35-year-old man, they could extrapolate. Let's see how extrapolation works on your graph.

M If needed, extend your line of best fit so that it goes through the *y*-axis and extends to 100 mL. Using your graph, predict the mass reading for 95.0 mL of salt water solution. Record your prediction in Table 2.

N Add solution to the graduated cylinder until there is 95.0 mL in the cylinder. Use the pipette if needed to get the volume exact. Measure and record the mass in Table 2.

O Calculate the percent error between your predicted value and the actual value. Record your answer in Table 2.

GOING FURTHER

Even the *x*- and *y*-intercepts of a curve of best fit often relate to characteristics of the system. The *y*-intercept is the value of the graph when the *x*-coordinate is zero.

7. What does the *y*-intercept on your graph represent?

_____

P Using your graph, predict the mass of the empty graduated cylinder. Record your prediction in Table 2.

Q Empty the graduated cylinder and dry it completely. Measure and record its mass in Table 2.

Name _____

R Calculate the percent error between your predicted mass and the actual mass. Record your answer in Table 2.

8. Why did you need to dry the cylinder completely?
   _____
   _____

9. Is your graph a reliable predictor of the empty cylinder's mass? Explain.
   _____
   _____
   _____
   _____

Table 1 DATA

TRIAL	VOLUME (mL)	MASS (g)
1		
2		
3		
4		

Table 2 TESTING PREDICTIONS

PREDICTION	GIVEN	PREDICTED	ACTUAL	% ERROR
1	mass (g)	volume (mL)	volume (mL)	
	140.00			
2	volume (mL)	mass (g)	mass (g)	
	95.0			
3		mass (g)	mass (g)	

Visual Data 19

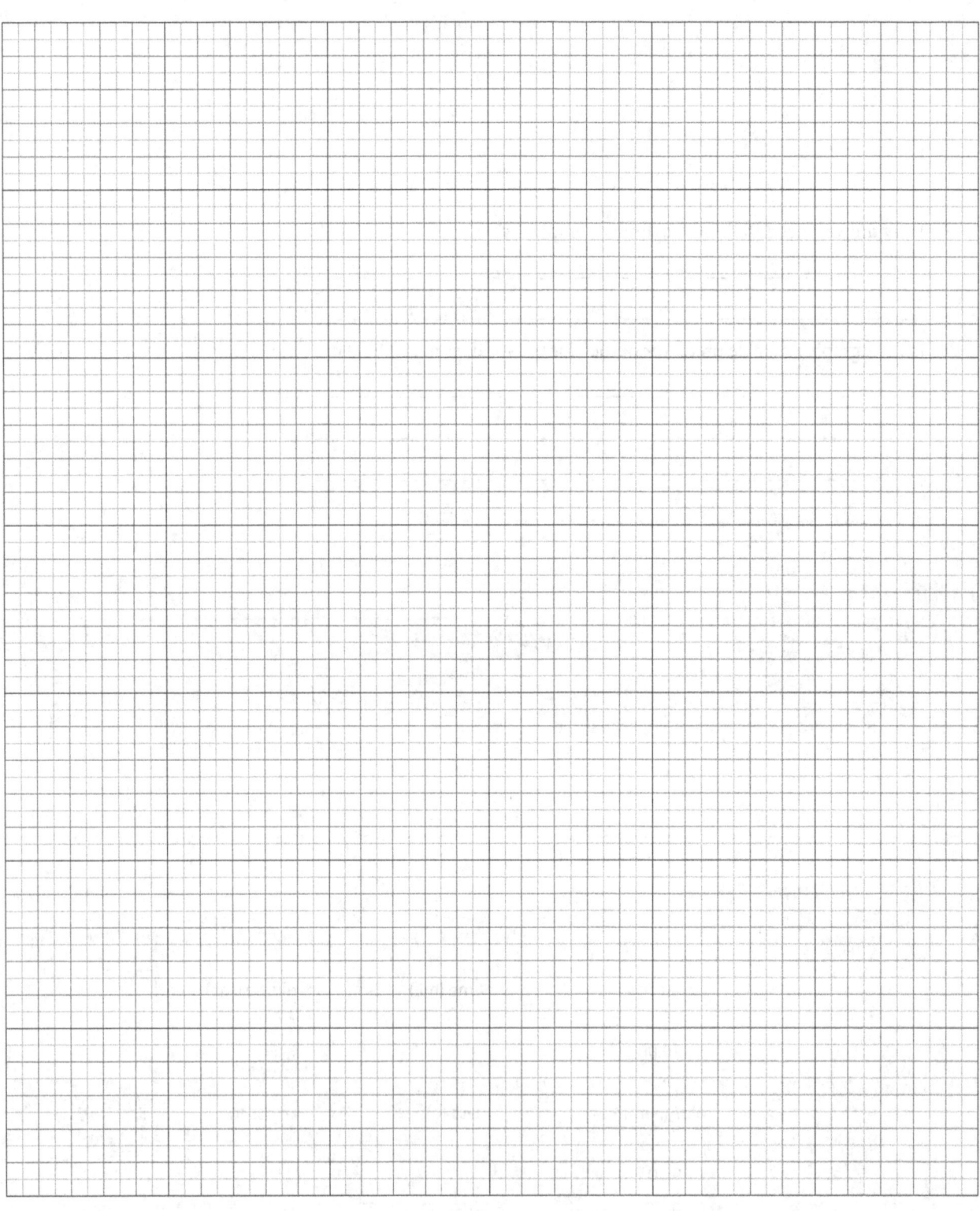

CHAPTER 2: MATTER

LAB 2A

HAS MASS, OCCUPIES SPACE

Modeling Matter

Some of the most basic ideas in science are the hardest to pin down. Physical science couldn't get along without the concept of matter, but it is very hard to define. In fact, scientists usually have to resort to listing the measurable properties of matter in order to define it. This results in an *operational definition*, a definition of what something does rather than what it is. We define matter as something that has mass and occupies space.

Measuring something's mass can present some unusual challenges. For example, how do scientists know that the earth has a mass of 5.972×10^{24} kg? After all, it's pretty hard to find a scale big enough for the earth! The answer lies in the second key property of matter—volume ("occupies space").

A British experiment in the 1770s attempted to determine the mass of the earth indirectly on the basis of its estimated volume. By taking measurements around a nearly symmetrical Scottish mountain they estimated the mass of the earth. You can use this same technique to estimate the mass of other objects that you cannot fit on a scale. In this lab activity, you will use model cars to estimate the mass of the real vehicles.

QUESTIONS

How are physical models created?

- How can someone create a model to estimate the mass of a large object?
- What assumptions are present in a scientific model?
- What possible consequences can assumptions have on a model?

Equipment

overflow can
metric ruler
graduated cylinder, 100 mL
car models (3)
paper cup
strong thread
goggles

PROCEDURE

Modeling a Car

As you collect data, use the appropriate rows of Tables 1, 2, and 3 to record your data.

A Obtain three model cars and record the type of each in Tables 1, 2, and 3.

B Use the ruler to measure the length of each model. Since actual cars are more conveniently measured in meters, not centimeters, convert your lengths to meters. Record these lengths in Table 1.

C Do some research on the car types of your models. Enter the lengths of these vehicles in Table 1. (Also find the actual masses of these vehicles and record the data in Table 3.)

D Calculate the scale of each model using the formula below. Enter your scales in Table 1.

$$scale = \left(\frac{length_{actual}}{length_{model}}\right)$$

Determining the Volume

Now that you know the scale of each of your models, you need to find out how much space each one occupies. This property of matter is called its *volume*. Because a car is rarely shaped like a cube, it's difficult to find its volume by measuring it. Instead, you will use the *displacement method* to find your models' volumes.

E Fill the overflow can nearly full with water. You will need to situate the can so that the graduated cylinder can be placed below the outflow spout.

F Using the paper cup, very carefully add water to the overflow can until it begins flowing from the spout. (Consider having another container in place to catch the excess water.)

G Attach a length of thread to one of the car models.

H Hold the graduated cylinder under the overflow can's spout so that it will catch the overflow.

I Slowly lower the model into the overflow can, catching the overflow in the graduated cylinder. When the model is fully submerged and the water has stopped flowing, remove the model from the overflow can and place the graduated cylinder on the table.

J Measure the volume of water in the graduated cylinder. Record the volume in Table 2. Empty the graduated cylinder.

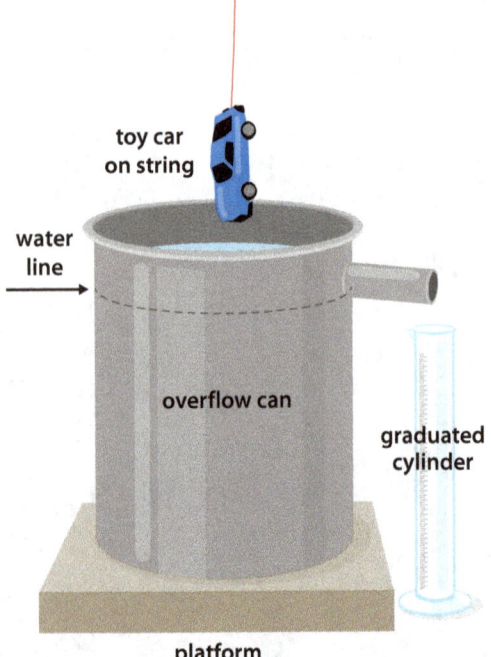

1. Why do you think that the volume of the overflow water represents the volume of the model car?

   _____

   _____

   _____

K   Determine the volume scale by cubing the length scale in Table 1. Record the volume scale in Table 2.

2. Why do you think that you had to cube the scale to get the scale factor for volume? (*Hint*: Think about calculating the volume of a box.)

   _____

   _____

   _____

L   Estimate the volume of the actual car by multiplying the model volume by the scale factor from Step K. Record your estimate in Table 2.

M   Repeat Steps F–L until you have determined the volume of all the cars.

Determining the Mass

Now you know the estimated volume of each car, but how does the model's volume relate to its mass? As you learned in Chapter 1, density is mass divided by volume, so you can calculate an object's mass if you know its density and volume.

N   Calculate the estimated mass of each car by using the density formula. Use 0.000 14 kg/mL as the density of an average car. It might surprise you that the density of a car (0.14 g/mL) is significantly less than water (1.0 g/mL). But if you think about it, much of the volume of a car is filled with air. Record the values in Table 3.

Evaluating Models

As you've already learned, no model perfectly matches what it represents. Models tend to be simplifications of the real thing, and they usually contain a few assumptions. In our case, we started with model cars to make an estimate of the mass of the real cars.

Scientists evaluate their models by listing the assumptions that they make and the possible weaknesses that the models might have. They might also think of ways to improve their models and make them more accurate. Let's evaluate our model along those lines.

3. What assumptions did we make right at the beginning of our modeling process? (*Hint*: We made it before we even did the first step.)

   _____

   _____

   _____

4. Where do you think the length data for real cars comes from?
   _____
   _____

5. If the average density of a car that you used to calculate the mass of the models is inaccurate, how would it affect the quality of your mass estimate?
   _____

O Calculate the percent error of each of your car mass estimates using the formula below. Record the percent error for the estimated mass of each type of car in Table 3.

$$\text{percent error} = \left(\frac{\text{mass}_{estimated} - \text{mass}_{actual}}{\text{mass}_{actual}}\right) 100\%$$

6. Which type of car had the smallest percent error?
   _____

7. Which type of car had the greatest percent error?
   _____

8. What do you think caused some of your estimates to be more accurate than others?
   _____
   _____
   _____

GOING FURTHER

9. How do you think the scientists in the 1770s used a method like this to determine the mass of the earth from the measurement of a Scottish mountain?
   _____
   _____
   _____

10. Why do you think the scientists in the 1770s wanted to use a symmetrical mountain?
    _____
    _____

11. The 1770s experiment concluded that the density of the earth was 4500 kg/m³. The modern accepted density of the earth is 5515 kg/m³. What does this indicate about the nature of modeling?
    _____
    _____
    _____

Name _____

Table 1 LENGTHS

TYPE	MODEL LENGTH (cm)	MODEL LENGTH (m)	ACTUAL LENGTH (m)	LENGTH SCALE

Table 2 VOLUMES

TYPE	MODEL VOLUME (mL)	VOLUME SCALE	ESTIMATED ACTUAL VOLUME (mL)

Table 3 MASSES

TYPE	ESTIMATED MASS (kg)	ACTUAL MASS (kg)	PERCENT ERROR

CHAPTER 2: *MATTER*

LAB 2B

SOMETHING OLD, SOMETHING NEW?

Detecting Physical and Chemical Changes

Every day scientists perform experiments and make observations. Observations are things that they detect with their senses or with instruments. For example, if a scientist notices that a substance changes color, that's an observation.

But observations alone aren't worth much. Scientists must use their observations to make *inferences*—the things that they conclude on the basis of their observations. If a scientist were to conclude that a color change indicates a chemical change, that would be an inference.

In this lab activity, you will make observations as you perform experiments. Then you will use those observations as evidence to make inferences about whether the experiment produced a physical or chemical change.

So how will you tell the difference between the two types of changes? You must look at your evidence. Observing one or more of the following phenomena often indicates that a chemical change has occurred.

A solid forms and separates from a liquid; it may or may not sink to the bottom of the container.

A gas appears in the form of bubbles.

A permanent color change occurs.

The substance's temperature changes.

Light or sound is produced.

QUESTIONS

How can I know whether a new substance forms when I mix two chemicals?

- What is the difference between a physical change and a chemical change?
- What are evidences that a chemical change has occurred?

But some physical changes include phenomena similar to those just stated. For example, the appearance of salt crystals in a slowly evaporating salt solution is a physical change. Boiling water, another physical change, produces gas and sound. Nonpermanent color changes can accompany physical changes, as can temperature changes. If a scientist wants more certain evidence that a chemical change really has happened, he must analyze the suspected new substance in a laboratory to confirm its identity.

PROCEDURE

Mixing Salt and Water

As you collect data, use the appropriate rows of Table 1 to record your data.

A Using the graduated cylinder, measure 25 mL of water and pour it into the Erlenmeyer flask.

B Place the thermometer in the flask and note the temperature of the water.

C Add 1/2 tsp of table salt to the water and swirl until the salt is dissolved.

D Observe what happened to the salt and water and record your observations by writing *Yes* or *No* in each cell of the *Salt and Water (Step D)* row of Table 1. Then decide whether the change was chemical or physical and record your conclusion in Table 1.

E Place the flask on the hot plate and turn the burner on high.

F After most of the water has boiled off, set the burner to low and continue heating until all the water has evaporated.

While the water is boiling, most of the lab group can continue on to the next procedure, but *at least one member must monitor the flask on the hot plate at all times*.

G As soon as the water has evaporated entirely, turn off the hot plate. Using the hot mitt, remove the flask from the hot plate and place it on the wire gauze.

H Let the flask cool completely and observe its contents, recording your observations and inference in the *Salt and Water (Step H)* row of Table 1.

I Rinse the flask with water.

Equipment

hot plate
graduated cylinder, 100 mL
Erlenmeyer flask, 250 mL
thermometer
plastic teaspoon
hot mitt
wire gauze
beaker, 100 mL
sodium chloride (NaCl), ½ tsp
white vinegar ($C_2H_4O_2$), 35 mL
baking soda ($NaHCO_3$), 1 tsp
modeling clay
red cabbage juice, 50 mL
milk of magnesia ($Mg(OH)_2$) solution, 10 mL
goggles

Mixing Baking Soda and Vinegar

J Use the graduated cylinder to pour 25 mL of vinegar into the beaker.

K Place the thermometer in the beaker and note the temperature of the vinegar.

L Add about 1 tsp of baking soda to the vinegar.

M Use the thermometer to detect any temperature change.

N Record your observations and inference in Table 1.

O Pour the beaker's contents down the drain and rinse the beaker with water.

Stretching and Compressing Modeling Clay

P Insert the thermometer into the clay and note the temperature.

Q Remove the thermometer and stretch and compress the clay for three minutes.

R Reinsert the thermometer to detect any temperature change.

S Record your observations and inference in Table 1.

Mixing Red Cabbage Juice and Milk of Magnesia or Vinegar

T Use the graduated cylinder to pour 10 mL of cabbage juice into the beaker.

U Rinse the graduated cylinder with tap water and then use it to add 10 mL of milk of magnesia to the cabbage juice.

V Record your observations and inference in Table 1.

W Flush the mixture down the drain with plenty of water and thoroughly rinse the beaker and graduated cylinder.

X Repeat Steps T–W but use 10 mL of vinegar in place of the milk of magnesia.

1. In which case(s) was there a color change?

   _____

   _____

2. In which case(s) was there a change in temperature?

   _____

   _____

3. In which case(s) was a gas produced?
   _____
   _____

4. Summarize your findings as a statement about how these characteristics can be used to determine whether a change is a physical or chemical change.
   _____
   _____
   _____
   _____

5. Explain why scientists generally collect a variety of observations when they're trying to establish whether a physical or chemical change has occurred.
   _____
   _____
   _____

6. Is there another explanation for any temperature change in the modeling clay? Is there a way to improve the procedure to eliminate this factor?
   _____
   _____
   _____

GOING FURTHER

7. You take a sickly plant and put it in the sunlight. A few days later the leaves are bright green. Did a physical or chemical change occur in the plant? Explain.
   _____
   _____
   _____

8. It's a hot day, and you notice resin oozing out of the boards of a picnic table. Is this a physical or chemical change? Explain.
   _____
   _____
   _____

Name _____

9. You are at the doctor's office, about to get a shot. The nurse swabs some alcohol on your arm. It feels cold for a few moments. Is this a physical or chemical change? Explain.

_____

_____

_____

10. You boil an egg on the stove. List some evidences that doing so causes a chemical change in the egg even though boiling is usually associated with physical changes.

_____

_____

_____

_____

Table 1

SUBSTANCE(S)	OBSERVATIONS			INFERENCE
	COLOR CHANGE?	TEMPERATURE CHANGE?	GAS PRODUCED?	TYPE OF CHANGE
Salt and Water (Step D)				
Salt and Water (Step H)				
Baking Soda and Vinegar				
Modeling Clay				
Cabbage Juice and Milk of Magnesia				
Cabbage Juice and Vinegar				

Something Old, Something New?

CHAPTER 3: THE ATOM

LAB
3A

THE FIERY TRIAL

Using Flame Tests

Nothing beats the soothing crackle of a log fire in the fireplace on a cold night. Wood fires usually glow red, orange, and yellow, but with the addition of a flame color kit, they can turn all the colors of the rainbow. A flame color kit usually contains packets of powdered mineral salts or pinecones that have been treated with the salts (you'll learn more about mineral salts in Chapter 10). When tossed into a fire, the salts burn, producing a variety of colored flames. The Danish physicist Niels Bohr used this property of mineral salts to deduce something about the structure of atoms—those tiny particles of matter that are too small to see directly.

QUESTIONS

Why do burning mineral salts produce different colors of light?

- Can flame tests be used to identify unknown chemicals?
- What can flame tests tell us about the structure of atoms?

PROCEDURE

Flame Tests

A Fill each of the beakers with water. One will be for unused wooden splints and the other for disposal of used splints after each trial. Place your wooden splints in one of the beakers.

B Light the laboratory burner. Adjust the burner's fuel and air mixture until you get a steady blue flame. If you've never used a laboratory burner before, wait for your teacher to instruct you. *Remember that someone must monitor the burner whenever it is in use.*

C Dip a fresh, damp wooden splint into the first salt sample.

Equipment

Bunsen burner
beakers, 250 mL (2)
burner lighter
damp wooden splints
mineral salts
goggles
laboratory apron

The Fiery Trial   33

D Place the tip of the splint in the hottest part of the flame. *Be sure to tilt the burner slightly so that no grains of salt fall into the burner.* Wave the splint back and forth in this area until the salt begins to glow and affect the flame color. *Note*: Do not let the splint itself catch fire. If it does, immediately dip it in the disposal beaker to extinguish it. Repeat the test again with a fresh splint.

E Record the color that the flame changes to in Table 1. Be as descriptive as possible. For example, if the flame changes to a reddish-orange, write "red-orange" rather than just "red" or "orange." Many of the colors that you'll see are similar but subtly different.

F Repeat Steps C–E until you've tested all but the last two samples (the ones marked "unknown"). Use a fresh splint for each test!

Testing Unknowns

The last two samples are not identified. However, they are actually identical to two of the samples that you've already tested. You'll use the flame test to try to figure out which ones they are. Chemists use flame tests to perform basic identifications of unknown chemicals.

G Repeat steps C–E using the two unknown substances. Once you've recorded your observations, use the data from your previous tests to try to identify the unknowns. Write the name of the salt that you believe each unknown sample to be in Table 1.

H Extinguish the burner. Dispose of all used splints in the trash.

1. Why do you want the flame to be a steady blue flame?

   _____
   _____
   _____

2. Aside from the obvious reason, why do you not want the splint to catch fire during the test?

   _____
   _____
   _____

ANALYSIS

3. Did each salt sample produce light of varying colors or only a single color?

   _____

4. Did any samples produce the same color? Or did each sample produce a unique color?

   _____

5. Do you think that the distinctive color is due to the first element (the metal) or the second element (the nonmetal) in each of the salts? Explain.

_____
_____
_____

You probably already know that atoms consist of a nucleus, containing protons and neutrons, surrounded by electrons. The colors produced by heated mineral salts are the result of electrons absorbing thermal energy and then reemitting that energy as light. The colors give us a clue about how electrons are arranged in atoms.

6. Early models of atoms hypothesized electrons moving randomly around an atom's nucleus. Do your results support that hypothesis? Explain.

_____
_____
_____

CONCLUSION

7. On the basis of the results from your flame tests, what can you conclude about the motion of electrons around an atom's nucleus?

_____
_____
_____

8. On the basis of the results from your flame tests, what can you conclude about the structure of each salt's atoms compared with the atoms of the other salts?

_____
_____
_____

GOING FURTHER

9. How can the flame test be used to help a chemist identify an unknown chemical?

_____
_____
_____

10. What is an obvious limitation of the flame test as a means for identifying unknown chemicals?

   _____
   _____
   _____

11. You may recall from your study of earth science that visible light consists of different wavelengths of electromagnetic energy. On the basis of that information, suggest a way to improve the reliability of the flame test for identifying unknown chemicals.

   _____
   _____
   _____
   _____

Table 1

SALT	COLOR
Calcium Chloride	
Copper Chloride	
Lithium Chloride	
Manganese Chloride	
Potassium Chloride	
Sodium Chloride	
Strontium Chloride	
Unknown 1:	
Unknown 2:	

CHAPTER 3: THE ATOM

LAB 3B

BIG TIME

Inquiring into Scale Models of Atoms

You've probably seen a model of a molecule like the one above. But was it a *scale* model? A scale model is one in which the dimensions of the model accurately reflect the real dimensions of the thing being modeled. For example, a ship model might be built in such a way that 1 cm of model represents 50 cm of real ship. This can be indicated as 1/50 or 1:50 scale. In a 1/25 scale model, 1 cm of model represents 25 cm of subject, so a 1/25 scale model is larger than a 1/50 scale model of the same subject.

Atoms, of course, are far too small to be seen with the unaided eye, which is why we often use models to illustrate them. But do the models we typically see accurately reflect the real dimensions of atoms? That's the question that you will examine in this lab activity.

QUESTIONS

What would the dimensions of an atom look like if the atom were enlarged to a size that we could easily see?

- Just how big are atoms?
- Do most depictions of atoms accurately show scale?
- How can the actual dimensions of an atom be shown using a scale model?

Big Time **37**

Equipment
none

Each helium atom has two protons and two neutrons, along with two electrons in its first energy level.

PROCEDURE

Collect Data

A You will be constructing a scale Bohr model (a nucleus with electrons at a fixed distance from the nucleus) of a helium atom (see left). The first thing you need to know is the size of the atom that you will be modeling. You can find this information on the internet. Try doing a search using the keywords "atomic radius" or "diameter of an atom." You'll need to determine the size of a helium nucleus as well.

B Once you have found the radii of the helium atom and its nucleus, you will need to determine a scale for your model. The scale you choose will depend in part on what you choose to represent the nucleus of your model atom. You do not need to accurately model the nucleus itself. You simply need any round object to represent the nucleus. Whatever you choose for this purpose must be large enough to be easily seen. The size of your nucleus will in turn determine how far away you will need to place your electrons.

C Finally, you will need to find a way to present your model in class. As you'll see, this might very well be the most challenging part of the activity!

DATA

1. What is the diameter of a helium atom?
   _____

2. What is the diameter of a helium nucleus?
   _____

3. How much larger in diameter is a helium atom compared with a helium nucleus? What effect will your answers to Questions 1 and 2 have on the size of your scale helium atom model?
   _____
   _____
   _____
   _____

BUILD IT!

Now that you know the relative dimensions you'll be working with, it's time to "build" your model. You'll need to get creative with this! Not only do you need to make your model, you will need to think of a way to present it in class as well. Begin by choosing an object to represent the nucleus of your helium atom. The diameter of this object will determine how far away you will need to place your electrons in order for your model to be properly scaled. (*Note*: Because the electrons are very small compared with the size of the nucleus, your electrons do not need to be to scale.) Be prepared to present and discuss your model in class.

4. How has this exercise in modeling helped you understand the nature of science?

_____
_____
_____
_____

GOING FURTHER

5. Find a resource that lists the atomic radii of all the elements. What do you discover about the radius of a helium atom compared with atoms of other elements?

_____
_____

6. What does your answer to Question 5 tell you about the size of your model compared with models of atoms of other elements at the same scale?

_____
_____

7. What does your model mainly consist of? What conclusion can you draw from this about the nature of matter?

_____
_____
_____
_____

CHAPTER 4: THE PERIODIC TABLE

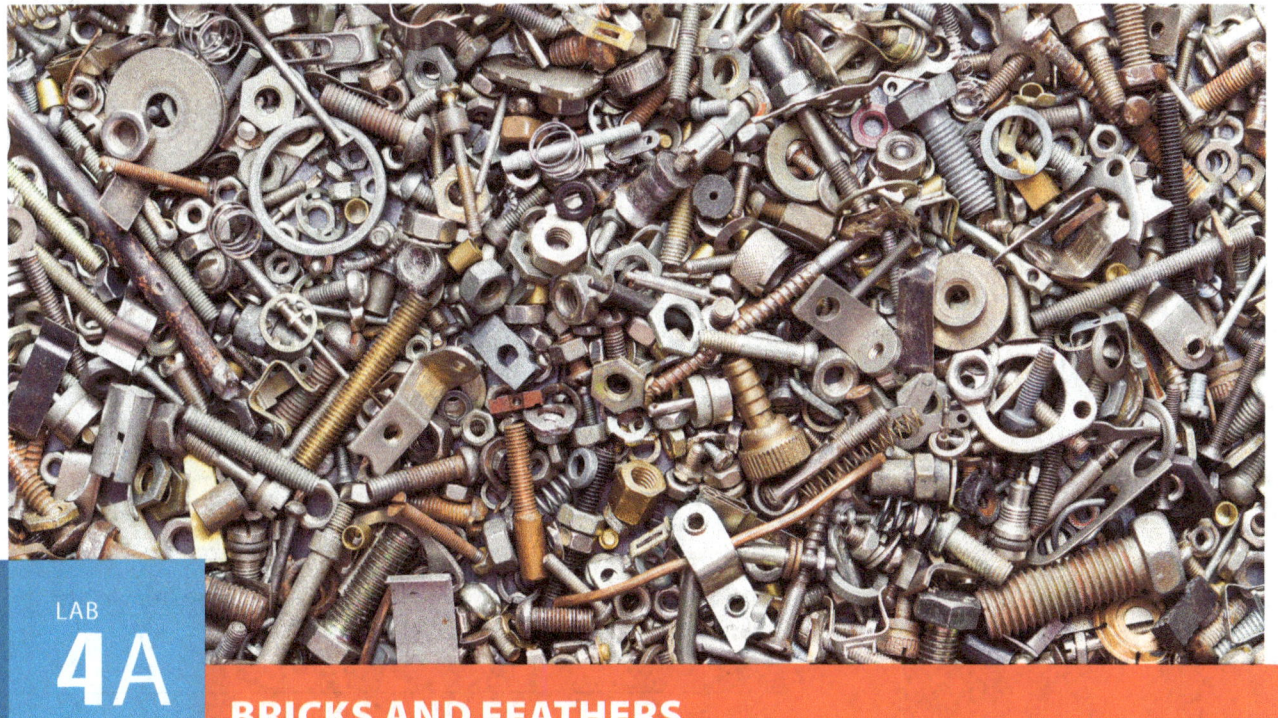

LAB 4A

BRICKS AND FEATHERS

Exploring Element Density in a Period

Which is heavier, a ton of bricks or a ton of feathers? Of course, they weigh the same, but the ton of feathers would take up a whole lot more space. This is because feathers are much less dense than bricks. It is this fact that makes this seemingly obvious query a trick question. If people aren't thinking, they may forget that a ton is a unit of weight, not volume, and answer that the ton of bricks weighs more.

In this lab activity, you will not have to weigh a ton of bricks or feathers, but you will be measuring the mass and volume of several different metal samples. You will calculate their densities and then develop a mathematical model regarding how density changes across a row of the periodic table.

QUESTIONS

Are the densities of elements in a row of the periodic table predictable?

- What are the densities of chromium, iron, and nickel?
- Is it possible to predict the density of cobalt?

PROCEDURE

Measuring Mass and Volume

As you collect data, use the appropriate rows of Tables 1 and 2 to record your data.

A Mass the sample of chromium and record the mass in Table 1.

B Use the overflow can to determine the volume of the chromium sample. Record the volume in Table 1. *Note*: If you need to review how to use the overflow can, see the instructions on page 22 in Lab 2A.

C Repeat Steps A and B for samples of iron and copper.

Equipment
laboratory balance
overflow can
graduated cylinder, 10 mL
chromium chunk
iron nails
copper shot
spreadsheet software
goggles
laboratory apron
nitrile gloves

1. On the basis of the periodic table, what do chromium, iron, and copper have in common?

   _____

Calculating and Graphing Volume

D Use the data that you collected to calculate the density of each element. Record the densities in Table 1.

E Enter the atomic numbers and densities of each element from Table 1 into a spreadsheet. Create a scatterplot. (*Note*: Place the atomic number on the *x*-axis.) Include a curve of best fit.

2. What is the dependent variable on your scatterplot? Explain.

   _____

   _____

   _____

3. What happens to the density of the metals as you move from left to right across the row?

   _____

   _____

   _____

F Using the curve of best fit, interpolate to predict the density of manganese (atomic number 25). To do this, find the point where the curve of best fit crosses the vertical line where the *x*-axis value is 25. The *y*-axis value at this point is the interpolated density of manganese. Enter your prediction in Table 2.

G Calculate the percent error for your prediction for the density of manganese and record the percent error in Table 2.

4. Assuming that a percent error less than 5% indicates an accurate estimate, is your estimate accurate?

   _____

H Repeat Steps F and G with cobalt (atomic number 27).

5. Is your prediction of the density of cobalt accurate? Explain.

   _____

   _____

6. If there was a significant difference in the accuracies of your two predictions, what do you think caused this difference?

   _____

   _____

   _____

7. Examine the graph on the right of all the Period 4 elements. Does its shape help explain any differences in the accuracies of your two estimates? Explain.

___

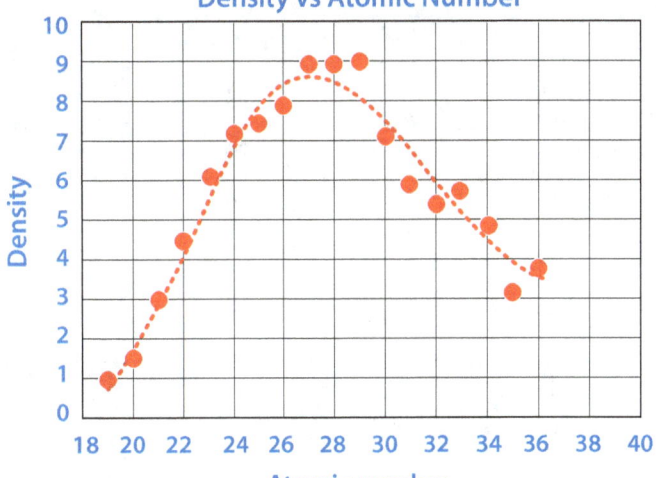

8. What fact about science does this illustrate?

___

GOING FURTHER

Knowing metallic density is useful in a variety of fields as diverse as airplane construction and fishing. Fishermen often put metal weights on their fishing line in order to cause the line and bait to sink. This puts the bait where the fish are more likely to be. The weights also keep the current from pushing the line downstream.

9. Given that the density of lead is 11 g/mL and the density of iron is 7.87 g/mL, why might lead be a better weight than iron? (Discount the fact that iron rusts in water while lead does not.)

___

10. The metal titanium is often used in airplane construction. Its density 4.51 g/mL. Why might titanium be a better material than iron for building airplanes? (Their strengths are roughly equal.)

___

Table 1 DATA

ELEMENT	ATOMIC NUMBER	MASS (g)	VOLUME (mL)	DENSITY (g/mL)
Chromium	24			
Iron	26			
Copper	29			

Table 2 PREDICTIONS

ELEMENT	ATOMIC NUMBER	PREDICTED DENSITY (g/mL)	ACCEPTED DENSITY (g/mL)	PERCENT ERROR
Manganese	25		7.3	
Cobalt	27		8.86	

CHAPTER 4: THE PERIODIC TABLE

LAB 4B

HOW WIDE IS AN ATOM?

Exploring Atomic Radii Trends

How many boxes can you fit on your closet shelf? Well, it depends on the size of the boxes. You can fit a lot more shirt boxes than large shipping boxes.

We know that atoms are inconceivably small. But the fact that they are so small often makes us think of them as all having the same size—tiny. But not all atoms are the same size. And this is important to consider when you start thinking about how much space atoms take up.

In this lab activity, you won't be estimating the size of atoms. Scientists have already done that for you. Instead, you will investigate atomic radii to determine whether there is a pattern in the atomic radii of atoms in a period or group.

QUESTIONS

Do atomic radii follow a periodic trend?

- What is atomic radius?
- Does atomic radius change predictably across periods or groups?

PROCEDURE

A Copy the atomic numbers and atomic radii data from Table 1 into your spreadsheet program. The unit for atomic radius is the angstrom (Å), which is equal to 0.1 nm.

B Create a scatterplot of the data from Table 1.

1. In which period (row) of the periodic table are these elements?

   _____

2. Do the atomic radii of these elements show a trend across the period? Explain.

   _____

Equipment
spreadsheet software

C Repeat Steps A and B with the data from Table 2.

3. In which period are these elements?
   _____

4. Do the atomic radii of these elements show a trend across the period? Explain.
   _____

5. On the basis of your analysis of the atomic radii of Periods 2 and 5, write a statement summarizing the trend of atomic radii across periods.
   _____
   _____

6. Create a hypothesis that explains the reason for this trend.
   _____
   _____
   _____
   _____

D Repeat Steps A and B with the data from Table 3.

7. In which group (column) are these elements?
   _____

8. Do the atomic radii of these elements show a trend down the group? Explain.
   _____

E Repeat Steps A and B with the data from Table 4.

9. In which group are these elements?
   _____

10. Do the atomic radii of these elements show a trend down the group? Explain.
    _____

11. On the basis of your analysis of the atomic radii of Groups 1 and 14, write a statement summarizing the trend of atomic radii down groups.
    _____
    _____

12. Create a hypothesis that explains the reason for this trend.
    _____
    _____
    _____

13. Combine your answers to Questions 5 and 11 to create a single statement summarizing the trend of atomic radii of the elements on the periodic table.

   _____
   _____
   _____

14. Silver has an atomic radius of 1.65 Å, while zinc's atomic radius is 1.42 Å. Predict a possible range for the atomic radius of copper.

   _____
   _____
   _____

15. This type of regular change in an element's characteristics across a period or down a group is called a *periodic trend*. What are some other characteristics that might have periodic trends?

   _____
   _____

Table 1

ELEMENT NAME	ATOMIC NUMBER	ATOMIC RADIUS (Å)
Lithium	3	1.67
Beryllium	4	1.12
Boron	5	0.87
Carbon	6	0.67
Nitrogen	7	0.56
Oxygen	8	0.48
Fluorine	9	0.42
Neon	10	0.38

Table 2

ELEMENT NAME	ATOMIC NUMBER	ATOMIC RADIUS (Å)
Rubidium	37	2.65
Strontium	38	2.19
Yttrium	39	2.12
Zirconium	40	2.06
Niobium	41	1.98
Molybdenum	42	1.90
Technetium	43	1.83
Ruthenium	44	1.78
Rhodium	45	1.73
Palladium	46	1.69
Silver	47	1.65
Cadmium	48	1.61
Indium	49	1.56
Tin	50	1.45
Antimony	51	1.33
Tellurium	52	1.23
Iodine	53	1.15
Xenon	54	1.08

Name _____

Table 3

ELEMENT NAME	ATOMIC NUMBER	ATOMIC RADIUS (Å)
Hydrogen	1	0.53
Lithium	3	1.67
Sodium	11	1.90
Potassium	19	2.43
Rubidium	37	2.65
Cesium	55	2.98
Francium	87	3.48

Table 4

ELEMENT NAME	ATOMIC NUMBER	ATOMIC RADIUS (Å)
Carbon	6	0.67
Silicon	14	1.11
Germanium	32	1.25
Tin	50	1.45
Lead	82	1.54
Flerovium	114	1.80

CHAPTER 5: BONDING AND COMPOUNDS

LAB 5A

THE SOLUTION TO A PROBLEM

Solubility and Chemical Bonds

Solutions are all around us. You might gargle with salt water to ease a sore throat, use contact solution to clean your contact lenses, or drink hot tea with honey. The seawater that covers over 70% of the earth is a solution. But not everything dissolves in water. The sand on the beach certainly doesn't dissolve. Our clothes are not soluble, or you wouldn't be able to wash them.

Water dissolves substances by the positive and negative ends of the water molecules pulling on charges within the material being dissolved. In this lab activity, you will test whether bond type (ionic versus covalent) can be used to predict solubility in water.

QUESTIONS

Does bond type indicate solubility?

- Do all substances dissolve in water?
- Is a substance's solubility related to its bond type?

Equipment
laboratory balance
Erlenmeyer flasks, 100 mL (3)
graduated cylinder, 50 mL
spatula
wax pencil
weighing paper
distilled water
sodium chloride (NaCl)
sodium bicarbonate (NaHCO$_3$)
potassium chloride (KCl)
vegetable shortening
vegetable oil
candle wax
copper(II) chloride (CuCl$_2$)
cellulose
goggles
nitrile gloves

PROCEDURE

1. On the basis of what you know about water's dissolving action, do you think that ionic or covalent substances will be more soluble? Write a hypothesis.

   _____
   _____

   A Use the wax pencil to label the three Erlenmeyer flasks 1, 2, and 3.

   B Use the graduated cylinder to add 50 mL of distilled water to Flask 1.

   C Use the laboratory balance to measure 3.0 g sodium chloride. Add the sample to Flask 1.

   D Repeat Steps B and C for the sodium bicarbonate (Flask 2) and the potassium chloride (Flask 3).

   E Swirl each of the flasks for 1 minute.

   F If all the material has dissolved, record a *Y* in Table 1. If some of the material remains, record an *N*.

   G Dispose of the solutions as directed by your teacher and rinse the flasks.

   H Repeat Steps B–G for the three covalent substances in Table 1.

2. State a claim about the solubility of ionic and covalent solutions. As always, support your claim with evidence from your data.

   _____
   _____

3. Does the data in Table 1 support your hypothesis? Explain.

   _____
   _____

4. Using the generalization that you created for Question 2, write a hypothesis about whether the two chemicals listed in Table 2 will dissolve in water.

   _____
   _____

   I On the basis of your hypothesis in Question 4, indicate in Table 2 whether you think each substance will be soluble.

   J Use the graduated cylinder to add 50 mL of distilled water to Flask 1.

   K Use the laboratory balance to measure 3.0 g copper(II) chloride. Add the sample to Flask 1.

52 Lab 5A

L Swirl the flask for 1 minute.

M If all the material has dissolved, record a *Y* in Table 2. If some of the material remains, record an *N*.

N Repeat Steps J–M for the cellulose.

O Dispose of the solutions as directed by your teacher and rinse the flasks.

5. Does the data in Table 2 support your hypothesis? Explain.

   _____
   _____

GOING FURTHER

If you live in or have ever visited the southern United States, you know that southerners love their sweet tea. But sugar is a covalent substance, and it dissolves quite well. How can that be?

SOLUBILITY OF COVALENT SUBSTANCES IN WATER		
Compound	Polarity	Solubility in Water
lard	nonpolar	insoluble
sugar	polar	soluble
isopropanol	polar	soluble

6. Can you figure out why sugar is able to dissolve in water? (*Hint*: See page 103 of your textbook.)

   _____
   _____
   _____

7. Each of the covalent substances that you tested in this lab activity are nonpolar covalent compounds. Use the information in the table above to write a generality about the solubility of polar and nonpolar covalent compounds.

   _____
   _____

Table 1

BOND TYPE	FLASK	SUBSTANCE	SOLUBILITY IN WATER
ionic	1	sodium chloride	
ionic	2	sodium bicarbonate	
ionic	3	potassium chloride	
covalent	1	vegetable shortening	
covalent	2	vegetable oil	
covalent	3	candle wax	

Table 2

BOND TYPE	FLASK	SUBSTANCE	PREDICTED SOLUBILITY IN WATER	EXPERIMENTAL SOLUBILITY IN WATER
ionic	1	copper(II) chloride		
covalent	2	cellulose solution		

CHAPTER 5: BONDING AND COMPOUNDS

LAB 5B
ELECTRIC LINES

Conductivity and Chemical Bonds

Have you ever been in a pool when the lifeguard orders everyone out because he has just heard thunder? Do you know why everyone should leave the pool? The reason is electric shock drowning.

Electric shock drowning is an accident caused by electricity while the victim is swimming. A swimmer can be killed by an electric shock, or he may be paralyzed by the shock and drown. While the electric current may come from lightning, it often comes from a faulty electrical component on a lighted dock or an electric boat. Electric shock drowning is a sad demonstration of the fact that water in lakes and streams conducts electricity.

But other substances do not conduct electricity very well. It is very difficult to conduct electricity through the air. In fact, lightning is one of the few examples of it occurring. Why do some solutions conduct electricity better than others? Could it have to do with the bond types of the dissolved substances? In this lab activity, you will test solutions of ionic and covalent substances to see whether there is a pattern between bond type and conductivity.

QUESTIONS

Which type of bond conducts electricity better?

- Is a substance's electrical conductivity related to its bond type?

Equipment

conductivity tester
beakers, 50 mL (3)
wax pencil
sodium chloride (NaCl) solution
sodium bicarbonate ($NaCOH_3$) solution
potassium chloride (KCl) solution
deionized water
sugar ($C_{12}H_{22}O_{11}$) solution
vegetable oil
copper(II) chloride ($CuCl_2$) solution
starch solution
goggles
nitrile gloves

PROCEDURE

1. Write a hypothesis comparing the conductivity of ionic and covalent solutions.

   _____

   _____

A. Use the wax pencil to label the three beakers 1, 2, and 3.

B. Pour approximately 15 mL of the sodium chloride solution into Beaker 1. Wipe the probe of the conductivity tester and lower it into the solution.

C. If the conductivity tester indicates conductivity, record a *Y* in Table 1. If the tester indicates no conductivity, record an *N*.

2. Look at the conductivity tester. Hypothesize about how it works.

   _____

   _____

D. Wipe off the probe of the conductivity tester.

E. Repeat Steps B–D for the sodium bicarbonate and potassium chloride solutions.

F. Dispose of the solutions as directed by your teacher. Rinse the beakers with deionized water.

G. Repeat Steps B–D for the three covalent solutions in Table 1.

H. Dispose of the solutions as directed by your teacher. Rinse the beakers with deionized water.

3. State a claim about the conductivity of ionic and covalent solutions. As always, support your claim with evidence from your data.

   _____

   _____

4. Does the data in Table 1 support your hypothesis? Explain.

   _____

   _____

5. Using the generalization that you created for Question 3, write a hypothesis about whether the two chemicals listed in Table 2 will conduct electricity.

   _____

   _____

I Indicate in Table 2 whether you think each substance will be conductive.

J Test the conductivity of the two solutions and record the results in Table 2.

6. Does the data in Table 2 support your hypothesis? Explain.

   _____

   _____

GOING FURTHER

7. Ionic compounds in the solid state do not conduct electricity. Why do you think this is so?

   _____

   _____

   _____

8. As you noticed above, deionized water does not conduct water well, if at all. If this is true, why is electric shock drowning a danger? On the basis of what you have learned today, suggest an explanation of the danger of water and electricity.

   _____

   _____

Table 1

BOND TYPE	FLASK	SUBSTANCE	CONDUCTIVITY
ionic	1	sodium chloride solution	
ionic	2	sodium bicarbonate solution	
ionic	3	potassium chloride solution	
covalent	1	sugar solution	
covalent	2	deionized water	
covalent	3	vegetable oil	

Table 2

BOND TYPE	FLASK	SUBSTANCE	PREDICTED CONDUCTIVITY	EXPERIMENTAL CONDUCTIVITY
ionic	1	copper(II) chloride solution		
covalent	2	starch solution		

CHAPTER 6: THE CHEMISTRY OF LIFE

LAB 6A

STICKY BUSINESS

Inquiring into Glues

You need to buy some glue for a school project. You walk into the store and are amazed by the number of different glues that are available. Why do you think there are so many types of glue?

Glues work through a combination of cohesive and adhesive forces. *Cohesive forces* are the forces that hold different particles of glue together, while *adhesive forces* are attractive forces between the glue particles and the surfaces being connected. Different glues are designed to work on different surfaces and in different conditions.

One of the most amazing uses of glue is to close wounds after surgery. Medical glues have improved wound healing and decreased the incidence of infections of these wounds.

In this lab activity, you will learn how one type of glue is made. You will then have the opportunity to improve how well the glue works by modifying its formulation.

QUESTIONS

How can I improve a casein glue?

- Why do we separate the milk into its components to make glue?
- How would different ingredients affect the performance of the glue?
- Which type of milk makes better glue?

Equipment

hot plate
laboratory balance
graduated cylinder, 100 mL
glass beaker, 400 mL
graduated cylinders, 25 mL (2)
stirring rod
wire gauze
filter funnel
Erlenmeyer flask, 250 mL
spatula
filter paper
pH paper
whole milk
white vinegar (CH_3COOH)
baking soda ($NaHCO_3$)
goggles
laboratory apron
nitrile gloves
hot mitt

PROCEDURE

What determines whether a glue is effective? Scientists in product testing must develop controlled experiments to evaluate the effectiveness of their products. As you work in this lab activity, you will need to assess the changes in the glue composition and production to determine whether they improved the glue's performance.

Designing Scientific Investigations

A Develop an experiment to test the effectiveness of glue.

B Have your teacher approve your procedures.

Planning/Writing Scientific Questions

C Using the 100 mL graduated cylinder, measure 100 mL of milk and pour it into the beaker.

D Using one of the 25 mL graduated cylinders, measure 20 mL of vinegar and pour it into the beaker.

The acid in the vinegar will cause the milk to separate into curds (solid) and whey (liquid). The curds include the protein *casein*. The separated casein can be used as an adhesive.

E Using the hot plate, heat the milk vinegar mixture over medium heat, stirring continuously. ***Any heat source must be monitored any time it is in use.***

F Once lumps start to form, turn off the hot plate. While wearing the hot mitt, remove the beaker from the hot plate. Place the beaker on the wire gauze and continue stirring until no more lumps form.

G Use the filter funnel, filter paper, and Erlenmeyer flask to separate the solid from the liquid.

H Once the filtration is complete, discard the filtrate (liquid) down the drain. Carefully squeeze any remaining liquid from the filter paper and solid.

I Use a spatula to transfer the curds (solids) back into the beaker.

J Using a clean 25 mL graduated cylinder, add and stir in enough water (10–15 mL) to the solids to make a smooth mixture (similar to white glue).

K Add 1 g of baking soda to the mixture in the beaker to neutralize the acid.

L Test the mixture with pH paper to confirm that the mixture is neutral (pH 7). If necessary, add more baking soda to reach neutrality.

M Following your glue testing procedures, test the performance of the glue that you made.

N Write a specific question or questions about this glue, its components, and its production that you could answer by collecting data. Revise the above procedures to modify the glue. Have your teacher approve your procedures.

Conducting Scientific Investigation

O Following your procedures for glue formulation and testing, collect the data to answer the question(s) that you wrote.

Scientific Argumentation

Scientific claims should state the claim and always be supported with evidence from experiments.

P Write a claim about your modified glue. As always, support your claim with specific evidence from your experiment.

GOING FURTHER

1. Why should a chemist who is a Christian consider working in materials science to improve glues?

   _____
   _____
   _____
   _____
   _____
   _____

CHAPTER 6: THE CHEMISTRY OF LIFE

LAB 6B
MILKING CHEMISTRY

Proteins in Food

In the desolate, frozen plains of Antarctica, a group of male emperor penguins huddle in the brutal winds and isolating darkness of polar winter. And yet, under the downy folds of each of these penguins lies a solitary egg, soon to hatch and reveal a chick to this seemingly inhospitable world. But where will the chick's first meal come from? All the emperor mothers are back at the coast, gorging on fish to recover from laying their eggs. Only the males remain behind to protect the eggs.

But these feathered fathers have a trick up the sleeves of their tuxedoes! Emperor penguins, flamingos, pigeons, and doves can produce something called *crop milk*. Though not the same as the milk that mammals produce, this milk is also high in protein, which is just what chicks need to get a good start in life. The milk comes from a lining inside the fathers' mouths that stores partially digested food. A father simply regurgitates this food to feed to his chick.

You've been learning about proteins—large molecules made of amino acids chemically linked together by *peptide bonds*. The major protein in cow's milk is casein, but it also contains the proteins *albumin* and *globulin*. Albumin is a protein also found in egg whites. In this lab activity, you will use a combination of chemicals to detect proteins in foods, including milk.

QUESTIONS

Which foods are a source of protein?

- How can I detect protein in foods?
- Why do I need protein?

Milking Chemistry 63

Equipment

laboratory balance
test tubes (7)
test tube rack
disposable pipette
mortar and pestle
beaker, 150 mL
graduated cylinders, 10 mL (2)
biuret solution
fresh milk
fermented milk
egg whites
distilled water
food to be tested
acetone
sand
goggles
laboratory apron
nitrile gloves

Keeping the Tubes Straight

The contents for each test tube:
1–biuret solution
2–fresh milk
3–fermented milk
4–egg whites
5–distilled water

The copper chelate that forms during the biuret test for proteins

PROCEDURE

The Protein Test

A Label five clean test tubes with the numbers 1–5. Arrange them in order in the test tube rack.

B Fill each test tube about one-third full with, in this order: biuret solution, fresh milk, fermented milk, egg white, and distilled water.

C Make a prediction about whether you think the substances in Tubes 2–5 will contain protein. Record your predictions in Table 1.

Biuret solution contains sodium hydroxide (NaOH)—a strong base. Bases are corrosive and therefore must be handled with care. If you spill any biuret solution on your skin, wash immediately and notify your teacher.

In the procedure below, you will test the contents of Tubes 2–5 for the presence of proteins by adding biuret test solution (Tube 1). Biuret solution contains sodium hydroxide (NaOH) and copper (II) sulfate ($CuSO_4$). The sodium hydroxide is needed because the interaction can occur only in an alkaline environment. The copper from the copper (II) sulfate shares electrons with the nitrogen atoms in the peptide bonds of the protein to form a structure called a *chelate* (see diagram below left). In the chelate, the electrons are shared by the nitrogen and copper atoms. The chelate makes the color change to violet or pinkish violet.

1. What kind of bond forms between the nitrogen and copper atoms?
   _____

2. Do you think that the biuret test could be used to compare relative amounts of protein in foods? Explain.
   _____
   _____

3. Is the biuret test a qualitative or quantitative test? Explain.
   _____
   _____

D Use a pipette to add 8 drops of biuret solution from Tube 1 to each of Tubes 2–5.

E Mix each tube by agitating it. Allow the test tubes to stand for 3–5 minutes.

F Record your observations of the biuret test in Table 1, including whether your results indicate the presence of protein.

Sometimes the biuret test produces a pinkish color change or a faint purple color. This happens when there are peptides present. So what's the difference between a peptide and a protein? Peptides involve fewer amino acids—as few as two. Proteins consist of several peptide chains that coil into complex shapes. They can involve hundreds or even thousands of amino acids. So proteins are chemically similar to peptides—just bigger!

4. What did you notice in the results of the tests of fresh milk and fermented milk?

   _____
   _____

5. Why do you think this happened?

   _____
   _____
   _____
   _____

Don't be Fooled!
Don't mistake the deepening of any blue color for a color change to violet. A blue color is a negative result for the biuret test for protein.

Testing Other Foods

Let's test some other foods for the presence of proteins.

- **G** For solid food, weigh out a 2 g sample and grind it into small pieces with a mortar and pestle.

- **H** Add about 5 mL of acetone and allow the mixture to settle for several minutes. Decant the liquid into the beaker and dispose of the used acetone in accordance with your teacher's instructions. *Acetone has a low vapor point, meaning that it vaporizes easily. Make sure that there are no ignition sources when using the acetone.*

- **I** Add 1–2 g of clean sand to the residue in the mortar and continue grinding the food as you gradually add 10–15 mL of distilled water. Grind thoroughly until you have a well-pulverized suspension.

- **J** Decant the suspension into a labeled test tube, leaving the sediment behind.

- **K** For liquid food, simply fill a labeled test tube about one-third full.

- **L** Use the pipette to add 8 drops of biuret solution.

- **M** Mix each test tube by agitating it. Allow to stand for 3–5 minutes.

- **N** Use the blank rows of Table 1 to record your additional observations of the biuret test.

Check the nutrition labels of the foods that you tested for the presence of protein.

ANALYSIS

6. Describe your success in detecting proteins using the biuret test.

   _____
   _____

7. Describe any observations with regard to variations in the color or the intensity of the color that you observed.
   _____
   _____

8. Compare the amounts of protein on the nutrition labels to the deepness of the purple color that you observed.
   _____
   _____

9. Record any patterns that you observed in the foods that demonstrated a positive biuret test.
   _____
   _____
   _____

GOING FURTHER

○ Do some research on the importance of protein in the human diet.

10. Use what you have learned from your research to describe the importance of protein in your diet.
    _____
    _____

11. Why is it important for a Christian to maintain a healthy diet?
    _____
    _____
    _____
    _____

Table 1

	PREDICTION	COLOR OBSERVED	PROTEIN PRESENT?
Fresh Milk			
Fermented Milk			
Egg Albumin			
Water			

CHAPTER 7: CHEMICAL REACTIONS

LAB 7A
SCIENCE FAIR REVISITED

Types of Chemical Reactions

At science fairs it's common to see the "volcano reaction." The display is created by mixing acetic acid (white vinegar) with sodium bicarbonate (baking soda). These two household substances combine to produce a foaming eruption of a mixture of a solid, a liquid, and a gas. It's a handy way to simulate an erupting volcano, particularly if you add some red food coloring to make the mixture look more like red-hot lava. The vinegar and baking soda reaction offers some very interesting insights into chemistry.

In this lab activity, we're *not* going to make a model volcano. Instead, we're going to study this reaction to see how it works. We're also going to use it as a springboard into several other chemical reactions. So let's get bubbling!

QUESTIONS

What happens during a chemical reaction?

- Can I classify a chemical reaction on the basis of its chemical equation?
- How can I tell whether a chemical reaction is taking place?
- How can I tell what is being produced during a chemical reaction?

Equipment

gas generator
beaker, 100 mL
plastic teaspoon
acetic acid ($HC_2H_3O_2$)
sodium bicarbonate ($NaHCO_3$)
distilled water, neutralized
pH indicator solution
limewater
goggles
laboratory apron
nitrile gloves

PROCEDURE

Examining the Volcano Reaction

Before we start mixing reactants, let's take a moment to study the volcano reaction so that we know what's actually going on. Acetic acid is a covalent organic compound with the formula $HC_2H_3O_2$. When you mix it in water, it ionizes into two ions, a hydrogen ion (H^+) and an acetate ion ($C_2H_3O_2^-$). Sodium bicarbonate is an ionic compound made from a sodium ion (Na^+) and a bicarbonate ion (HCO_3^-). When you dissolve it in water, it dissociates into its component ions.

It turns out that the volcano reaction is actually two separate reactions. The second one occurs almost immediately after the first one. The first reaction is as follows.

$$HC_2H_3O_2 + NaHCO_3 \rightarrow NaC_2H_3O_2 + H_2CO_3$$

Two products result from this reaction—sodium acetate and carbonic acid.

1. What kind of reaction is this? Explain.

   _____
   _____
   _____

2. How do you know that the equation given above is balanced?

   _____
   _____
   _____

The second reaction involves the second product of the first reaction—carbonic acid. Under some conditions, this compound is stable, but in most cases the carbonic acid breaks down into water and carbon dioxide (CO_2). The second reaction is shown below.

$$H_2CO_3 \rightarrow H_2O + CO_2$$

3. What kind of reaction is this? Explain.

   _____
   _____

4. Express the volcano reaction in a single chemical equation.

   _____

Turning Sour

Now that we understand the volcano reaction, let's use one of its products—carbon dioxide—as the reactant in another chemical reaction. Since we usually encounter carbon dioxide as a gas, we'll use a piece of equipment called a *gas generator* (see next page) to help us use carbon dioxide in that state for a reaction.

A Remove the top from the gas generator and fill the bottle with acetic acid to a depth of 2 cm (about 75 mL).

B Fill the beaker with 50 mL of distilled water. Add about 20 drops of pH indicator solution to the water and gently swirl the beaker. The water should have a definite color. If not, add another 10 drops.

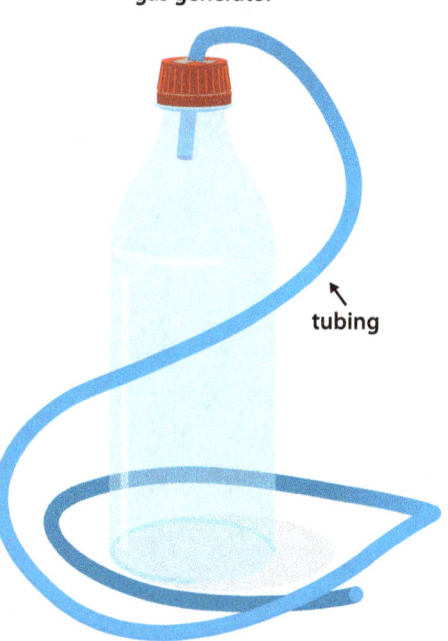

gas generator

tubing

By definition, distilled water should be nothing more than water molecules. We sometimes say that it's "neutral" since it doesn't contain ions from other substances. The pH indicator solution is a chemical designed to show the presence of compounds (called acids) in the water. You'll learn a lot more about acids in Chapter 10. For now, it's sufficient to know that as long as the water is pure, the indicator solution will be a light green color. If the water contains an acid, the indicator will turn yellow.

C Fold a sheet of notebook paper in half and then unfold it. Measure out one spoonful of sodium bicarbonate along the fold line.

D Hold the notebook paper in a V shape and use it to pour the sodium bicarbonate into the gas generator. Quickly cap the bottle.

E With one hand, gently agitate the gas generator. With the other hand, hold the generator's tube in the distilled water so that the carbon dioxide from the reaction bubbles through the liquid.

F When the reaction in the gas generator stops, remove the tube from the beaker and examine the distilled water.

5. Describe what happened to the pH indicator when you bubbled carbon dioxide through the distilled water.

   _____

6. What can you conclude from the pH indicator's behavior?

   _____
   _____
   _____

The following equation describes the chemical reaction that you've just produced.

$$H_2O + CO_2 \rightarrow H_2CO_3$$

As the carbon dioxide bubbles through the water, a molecule of carbon dioxide combines with a molecule of water to form a molecule of a new compound—carbonic acid. Carbonic acid in solution is partly responsible for the tang of sodas. You've already seen carbonic acid in the previous section. It was a short-lived product of the volcano reaction.

7. What kind of reaction is this? Explain.

   _____
   _____

Science Fair Revisited

G Gently stir the carbonic acid solution for several minutes, observing the color as you do.

8. What happens to the color of the indicator? What does this tell you about the pH of the solution?

   _____

   _____

9. Suggest an explanation for what you observed for Question 8.

   _____

   _____

   _____

Clouding Up

You've just seen how carbon dioxide reacts with water to produce carbonic acid. We're going to go a step further and react carbonic acid with limewater, a solution of an ionic compound called *calcium hydroxide* ($Ca(OH)_2$).

H Pour the contents of the gas generator down the sink, rinse the bottle with distilled water, and refill it with acetic acid to a depth of 2 cm (about 75 mL). Pour the contents of the beaker down the sink and rinse it out thoroughly.

I Fill the beaker with 50 mL of limewater.

J Measure out another spoonful of sodium bicarbonate onto the sheet of notebook paper.

K Hold the notebook paper in a V shape and use it to pour the sodium bicarbonate into the gas generator. Quickly cap the bottle.

L Cup the bottom of the gas generator in one hand and gently agitate it. With the other hand, hold the generator's tube in the limewater so that the carbon dioxide from the reaction bubbles through the liquid.

M As soon as the limewater has reacted, remove the tube from the beaker and examine the limewater against a bright light.

N Set the beaker aside where it will be undisturbed for a few minutes.

10. Describe the limewater after you've bubbled carbon dioxide through it.

    _____

    _____

From the previous experiment, we know that bubbling carbon dioxide through water gives carbonic acid (H_2CO_3). So it's reasonable to assume that the reaction we've just observed is a reaction between carbonic acid and calcium hydroxide.

When the calcium hydroxide dissolves in water, it dissociates into a calcium ion (Ca^{2+}) and two hydroxide ions (OH^-). The carbonic acid ionizes into one hydrogen ion (H^+) and one bicarbonate ion (HCO_3^-). The hydrogen ion combines with one of the hydroxide ions to form a water molecule.

$$Ca^{2+} + 2OH^- + H^+ + HCO_3^- \rightarrow Ca^{2+} + OH^- + HCO_3^- + H_2O$$

The bicarbonate ion then ionizes into another hydrogen ion (H+) and a carbonate ion (CO_3^{2-}). The hydrogen and the remaining hydroxide ion form a second water molecule.

$$Ca^{2+} + OH^- + H^+ + CO_3^{2-} + H_2O \rightarrow Ca^{2+} + CO_3^{2-} + 2H_2O$$

Finally, the calcium and carbonate ions combine.

$$Ca^{2+} + CO_3^{2-} + 2H_2O \rightarrow CaCO_3 + 2H_2O$$

The following equation describes the overall reaction.

$$Ca(OH)_2 + H_2CO_3 \rightarrow CaCO_3 + 2H_2O$$

11. What kind of reaction is this? Explain.

12. Describe the solution in the beaker after it has sat undisturbed for a few minutes.

13. What is the name of the precipitate?

14. What did you feel when you cupped the gas generator in your hand during the reaction?

GOING FURTHER

15. Does the volcano reaction give off energy? Support your answer.

CHAPTER 7: CHEMICAL REACTIONS

LAB
7B

IT'S IN THE BAG

Inquiring into Chemical Reactions

If you mix two chemicals together and a reaction occurs, it's pretty easy to tell which chemicals are reacting. But what if you mix *three* chemicals together and a reaction occurs? Could you easily identify which two chemicals of the three cause the reaction? And what if all three are responsible for the reaction? What if there is more than one effect? It can get confusing! In this lab activity, you will design and carry out a procedure for determining which chemicals are responsible for the observed evidences of a chemical reaction.

The three chemicals that you will use in this activity are calcium chloride, sodium bicarbonate, and phenol red. Calcium chloride is commonly used as a desiccant, a chemical that absorbs excess moisture. You may recognize sodium bicarbonate as baking soda. Pure phenol red exists as a solid crystal, but it is normally used in an aqueous solution, as in this lab activity.

QUESTIONS

How can I tell which reactants cause which effects?

- How can I tell whether a chemical reaction has occurred?
- How can I control an experiment?

Equipment

beakers, 100 mL (2)
beakers, 50 mL (2)
measuring spoon, 1 tbsp
measuring spoon, 1 tsp
disposable pipettes, 3 mL (2)
resealable sandwich bags
calcium chloride ($CaCl_2$), about 50 mL
sodium bicarbonate ($NaHCO_3$), about 50 mL
phenol red, 0.02% solution, about 40 mL
distilled water
goggles
laboratory apron
nitrile gloves

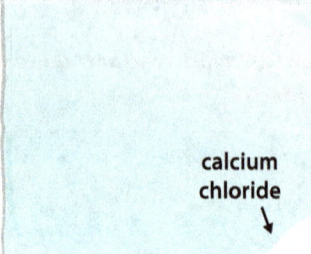

PART 1: OBSERVING THE REACTION

A Label one of the 100 mL beakers "calcium chloride" and the other "sodium bicarbonate." Fill each beaker about half-full with the correct chemical.

B Fill one of the 50 mL beakers with about 40 mL of phenol red and the other with the same amount of distilled water.

1. Describe the physical properties of the calcium chloride, sodium bicarbonate, and phenol red.

   _____

   _____

To prevent accidental contamination, each of the measuring spoons and pipettes that you use for the next steps should be used for only *one* chemical each.

C Place 1 tbsp of calcium chloride into one corner of a plastic bag (see below left).

D Place 1 tsp of sodium bicarbonate into the other corner of the bag. Lay the bag flat on your desk. Remove as much air from the bag as you can without mixing the chemicals.

E Fill a pipette with phenol red, and while keeping the bag flat, dispense the liquid into the mouth of the bag. Repeat this two more times, then seal the bag. Again, try to keep as much air as possible out of the bag.

F Pick up the bag and allow the chemicals to mix thoroughly. Carefully observe any reaction that occurs.

2. Describe any changes that you observe taking place in the bag.

   _____

   _____

   _____

PART 2: IDENTIFYING THE CULPRITS

As you observed, several things happened when the chemicals in Part 1 were mixed.

Writing Scientific Questions

G Look back at your observations. Write a series of questions about those observations that you could answer by conducting controlled experiments.

Designing Scientific Investigations

H As you think about your procedure, remember the scientific method and how experiments are designed. Think about what information you will need in order to answer your questions. Identify the variables involved in the Part 1 reaction and consider how you will treat them during your investigation.

I Write your procedures so that another scientist could recreate your laboratory experience.

J Have your teacher approve your procedures prior to beginning your investigation.

Conducting Scientific Investigations

K Carry out your investigation. Be sure to carefully record your observations. A data table is a good way to track your progress and present your findings. Obtain more of the reactants from your teacher if needed for additional tests.

Developing Scientific Models

L Analyze your data to answer the questions that you wrote in Step A.

Scientific Argumentation

3. In the space provided, summarize your findings. On the basis of these findings, make a claim for each question that you wrote in Step G. Support your claim with evidence from your experiment.

_____
_____
_____
_____
_____

CHAPTER 8: NUCLEAR CHANGES

LAB 8A
FLIPPING OUT

Modeling Radioactive Decay

When you hear the word "uranium," you probably don't think of colored glass. But in the nineteenth and early twentieth centuries, some glassware was colored with naturally occurring *uraninite*. This type of glassware, often called *uranium glass*, typically had a greenish tint.

Of course, uranium is much more famous for its use in nuclear power plants and nuclear weapons because of its radioactivity. (Its radioactivity is also why it is no longer used in dinnerware!) As scientists studied radioactivity, they learned about different kinds of radioactive decay. The most common isotope of uranium, U-238, goes through a long and complicated decay process before it reaches a stable isotope of lead. It is no coincidence that natural uraninite ore is always found mixed with traces of thorium, radium, lead, and helium since they are all produced by this process. The first three of these are different *daughter nuclides* that appear at different stages of uranium's decay process. The helium (alpha particles) are produced by alpha decay, one of uranium's main methods of radioactive decay.

One of the ways that scientists classify radioactive elements is by half-life, the time that it takes for 50% of the atoms in a sample to decay into daughter nuclides. Half-lives can be extremely long (billions of years) or extremely short (10^{-23} s). Understand that half-life refers to the probable behavior of a group of atoms in a sample, and not to knowledge about the individual atoms. There is no way of predicting whether a specific atom will decay.

This lab activity explores the concept of half-life. You will begin by taking a quick look at probability, after which you will move on to making a model of radioactive decay.

QUESTIONS

How can probability predict how long a sample remains radioactive?

- What is radioactive decay?
- How can radioactive decay be modeled?

Flipping Out 77

Equipment
pennies (32)
small box with lid
colored pencils (2)

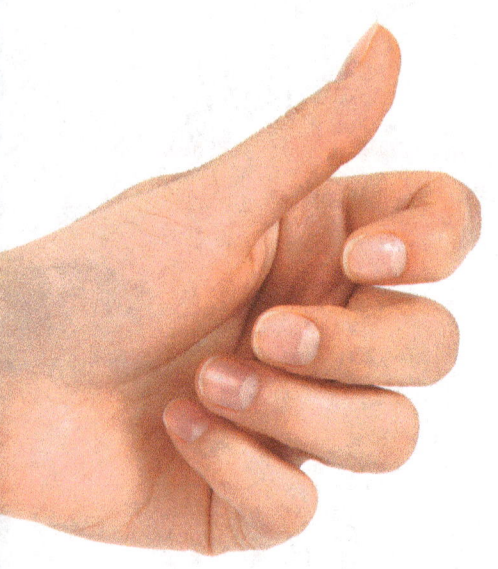

PROCEDURE

If you've ever flipped a coin, you've explored basic probability. Mathematicians and scientists use this branch of mathematics to help them know how likely it is that something will happen. Probabilities are normally reported as a decimal value between 0 and 1. Throughout this lab activity, we will use normal percent to report how likely events are. Probability math can be pretty complex, but there are also many simple applications. The coin toss is a good example.

1. When you flip a coin, what are the possible outcomes, including the percent likelihood for each outcome?

   _____
   _____

2. If you toss a coin 100 times, about how many times do you expect it will land heads up?

   _____

3. If you toss the same coin another 100 times, will the results be exactly the same as before? Explain.

   _____
   _____

4. If you toss a coin five times and it comes up heads each time, what is the likelihood that it will come up heads on the next toss? Explain.

   _____
   _____

Radioactive decay follows the same basic principle as the coin toss. Each radioactive atom has a 50% likelihood of decaying during one half-life. For example, the radioactive gas radon-222 has a half-life of 3.8 days. If you start with 100 atoms of radon, each atom has a 50% likelihood of decaying in the next 3.8 days. At the end of this time period, *approximately* 50 of the atoms will still be radon-222, while the remainder will have decayed to become polonium-218.

5. Assuming that you have 50 radon-222 atoms left after the first half-life, what do you expect to happen to those atoms during the next half-life?

   _____
   _____

Modeling Radioactive Decay

Thinking about radioactive decay as percentages isn't very easy to visualize, so let's do something that's pretty familiar to you by now. Let's make a model of the radioactive decay process.

A Place 32 pennies—representing atoms—in the box.

B Predict how many of these atoms will remain undecayed after one half-life. Record your prediction in the appropriate row of the *Trial 1 Predicted Remaining* column of Table 1.

C Cover the box and shake it up and down briefly but vigorously.

D Open the box and remove the pennies that landed tails up.

6. If the pennies represent radioactive atoms and we are removing the pennies that land tails up, what do you think these pennies represent?

_____

7. What does the shaking of the box represent?

_____

E Count the pennies that landed heads up and record them in the appropriate row of the *Trial 1 Actual Remaining* column of Table 1.

F Repeat Steps B–E until all the pennies have been removed from the box.

8. Were your predicted values accurate? Explain.

_____
_____
_____

G Plot your data in the graphing area on page 82.

9. What does the *x*-axis represent?

_____

10. What does the *y*-axis represent?

_____

H Repeat Steps A and C–G for Trials 2–5 so that you have five complete sets of data points.

I On your graph, draw one single smooth curve of best fit for all the points.

11. Describe the graph's behavior as the number of half-life time periods increases.

_____
_____

12. Was the data from the five trials a perfect match?

    _____
    _____

13. Theoretically, what is the likelihood of any individual atom decaying during each half-life?

    _____

J On the basis of your answer to Question 13, fill in the *Theoretical* column of Table 1 and plot it on your graph, using a different colored pencil.

14. How similar was the curve of best fit for your five trials to the theoretical graph?

    _____
    _____

15. From your experimental data, what is the mean and median number of half-lives before all the pennies had decayed? (Round to the nearest whole number.)

    _____

16. According to the mean that you calculated in Question 15, how long will it take for thirty-two atoms of the dangerous isotope cobalt-60 to decay completely to the harmless isotope nickel-60? Its half-life is 5.27 years.

    _____

17. In a sample of 32 atoms, what is the theoretical number of half-lives before the sample has decayed entirely?

    _____

18. If the sample is doubled so that it has sixty-four atoms, what is the theoretical number of half-lives before the sample has decayed entirely? Explain.

    _____

GOING FURTHER

One very popular form of nuclear medical imaging uses the radioactive isotope technetium-99m, which emits gamma rays as it decays and has a half-life of six hours. A medical technician injects the patient with a small amount of Tc-99m. As the technetium circulates in the patient's bloodstream, gamma rays pass from inside the patient to the outside world. A medical ray camera surrounding the patient detects the radiation, which a computer uses to create an image of the body system of interest. It's a bit like taking an x-ray from the inside out.

19. Using the data from your radioactive decay model, how long do you think it will take a thirty-two-atom sample of Tc-99m to decay?

_____

20. What problem does using an isotope with a 6-hour half-life cause for the companies that produce it as well as for the medical facilities that use it?

_____

_____

_____

Nuclear medicine facilities solve this problem by using another kind of nuclear decay. Instead of buying Tc-99m, they use a simple machine called a *technetium generator* to produce the required isotope when they need it.

A technetium generator (called "technetium cow" or "moly cow" by medical technicians) contains molybdenum-99. This isotope has a half-life of 66 hours. As it decays, it turns into Tc-99m. Technicians milk the generator by running a saline solution through the isotope to collect Tc-99m particles.

21. A technetium generator produces useful amounts of Tc-99m for about 3 half-lives of molbdium-99. How often would a nuclear medicine facility have to order a new generator?

_____

Several characteristics of technetium-99m make it fairly safe if used properly. Its short half-life means that it produces gamma rays for only a few days at most. It takes about a week for the body to release all the technetium atoms through urination. As the nuclei decay, they produce technetium-99, which emits beta radiation, but its half-life of 211 000 years means that it is extremely unlikely that more than one or two nuclei will decay before it is released from the body.

22. Describe a balance between using radioactive material to provide lifesaving medical diagnostics and minimizing exposure to harmful radiation.

_____

_____

_____

_____

_____

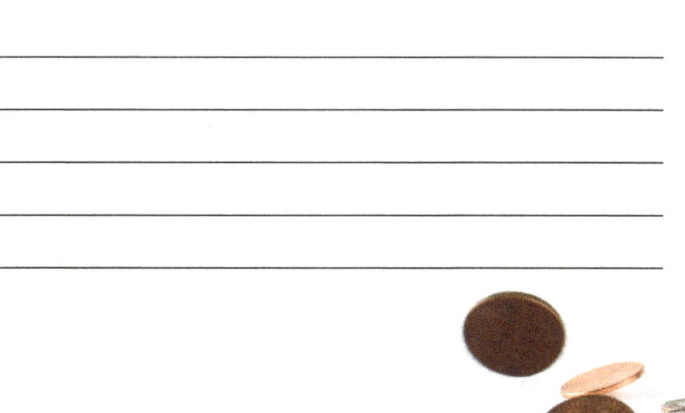

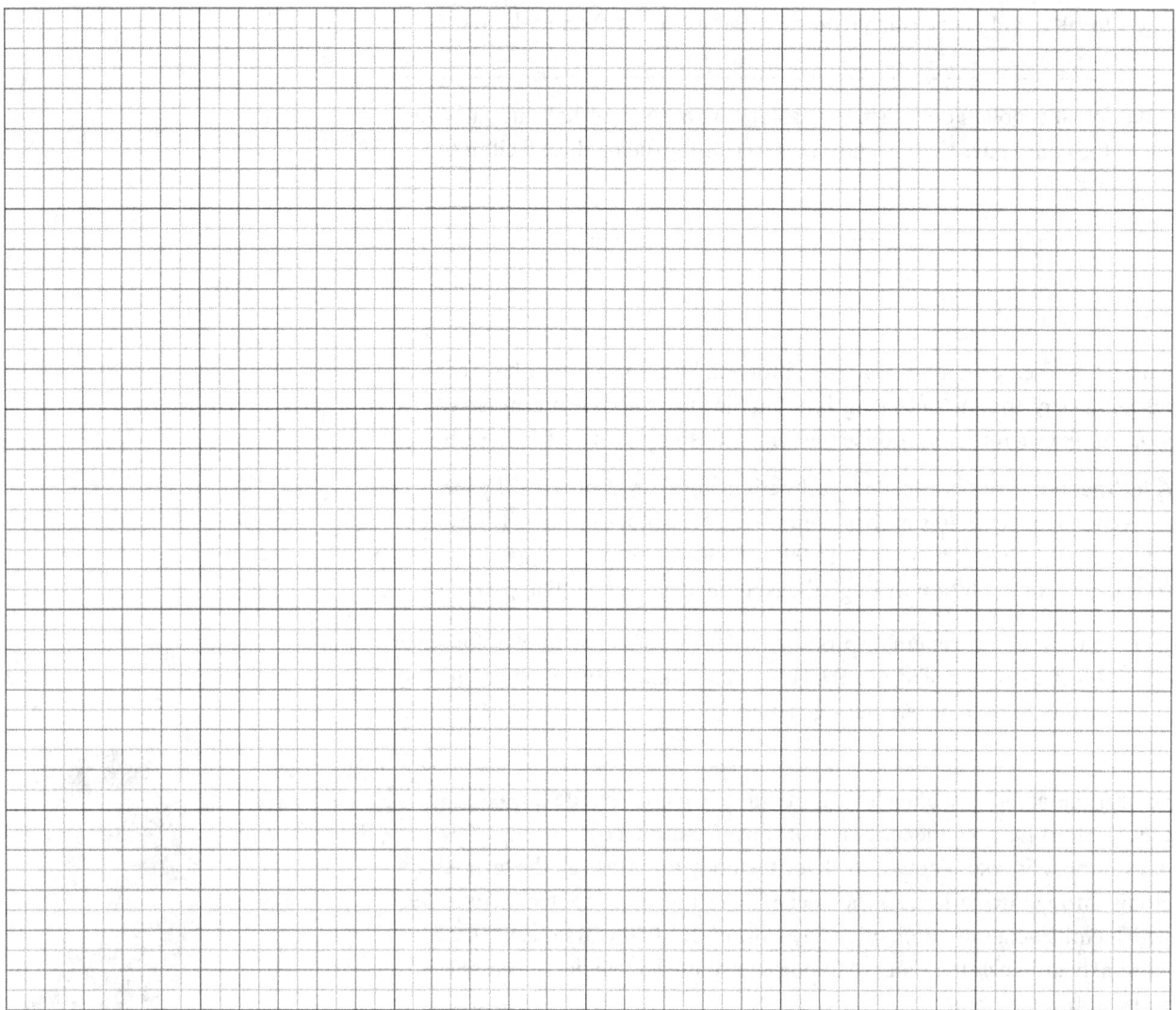

Name _____

Table 1

Half-Lives	TRIAL 1 Predicted Remaining	TRIAL 1 Actual Remaining	TRIAL 2 Actual Remaining	TRIAL 3 Actual Remaining	TRIAL 4 Actual Remaining	TRIAL 5 Actual Remaining	Theoretical
0	32	32	32	32	32	32	32
1							
2							
3							
4							
5							
6							
7							
8							
9							
10							
11							
12							
13							
14							
15							

Flipping Out

CHAPTER 8: NUCLEAR CHANGES

LAB 8B

RADIOACTIVE!

Exploring Radiation Dose

The idea of radiation exposure frightens many people. And while radiation can have a serious health impact, most of us are exposed to some radiation every day with no ill effects. Much of this radiation is background radiation from the sun or from radioactive minerals in the soil. Health problems from radiation increase with dose. A larger amount of radiation, especially repeatedly, will lead to severe health issues. But as long as the amount of radiation that a person receives remains below a certain level, the impact is probably minimal.

QUESTIONS

What is my annual radiation exposure?

- How are people exposed to radiation in their daily lives?
- What are common sources of radiation?
- What lifestyle choices increase my radiation exposure?

PROCEDURE

1. What do you think is the source of the largest amount of radiation that the average person is exposed to?

   _____

   A Use the internet to access a radiation dose calculator.

2. According to the radiation dose calculator, what are some of the categories of radiation sources?

   _____

   _____

Equipment

computer with internet access

Below are biographies of five fictional people. Read about each one, and then go on to the procedure steps that follow.

1. *Abner Brown*

Abner lives near the Gulf of Mexico in Mississippi. His brick house is located 500 feet above sea level, and he worries about radiation exposure because he lives 25 miles from a nuclear power plant. Since he is retired, he travels a good bit, spending 100 hours in the air this past year and flying nearly 45,000 miles (with luggage scanned on each trip). But he also enjoys equipping his house with the latest safety equipment, including smoke detectors, of course. He did have a heart scare this year, necessitating a cardiac CT scan.

2. *Nemo Waterman*

In spite of his name, Nemo prefers mountains and desert and is an avid hiker. He lives in the Colorado Plateau region of northeastern New Mexico at 5300 feet in elevation. Because he likes Spanish Colonial architecture, his house is built of adobe and equipped with smoke detectors. He hates to fly, so he avoids airports and cities if he can manage it. Because his house is nowhere near a power plant or even a power line, he generates his own electricity using solar power. But his life has its hazards. He fell earlier this year while hiking and required a chest and pelvis x-ray to determine the extent of his injuries.

3. *Rebecca Batissac*

Rebecca enjoys her life as a US military doctor, specializing in combat trauma. In her current assignment, she lives in York, PA (elevation 466 feet) in a wood-frame house equipped with smoke detectors. York is within 50 miles of both nuclear power and coal-burning power plants. In her role, she has flown 500 hours and 200,000 miles in the past year. Although she is military personnel, she is not exempt from the luggage scans in airports.

4. *Sidney Jug*

Sidney is a retired lawyer and history buff who, his friends say, loses his head over the French Revolution. His historic stone house in South Carolina is located at sea level on a sheltered cove on the coast. He has updated the house, including adding smoke detectors. He travels a good bit, and flew 100 hours and 50,000 miles this past year. His collection of historical artifacts raises eyebrows whenever his luggage is scanned at airports, but being a lawyer, he knows to make sure that his paperwork is in order. He lives about 40 miles from a coal-fired power plant.

5. *Artemis Vernon*

Artemis owns a landscaping business in her small community in Tennessee. Surrounded by mountains, her town sits at 1200 feet, and the Tennessee Valley Authority produces the town's power through hydroelectric rather than coal or nuclear. Her log cabin in the woods includes smoke detectors, and she and her husband occasionally fly to Northumberland, New York, to see their daughter. When they do, they don't take any luggage since their daughter's house has a separate furnished apartment for them. Because of her age, Artemis has yearly mammograms, and she needed a dental x-ray during her most recent routine dentist visit.

B Enter the information for each fictional person into the radiation dose calculator.

C Record each one's estimated annual radiation dose in Table 1.

D Record the source of the largest single dose of radiation for each individual in Table 1.

E Record the amount of the largest single dose of radiation for each individual in Table 1.

3. The average annual radiation dose of a person in the United States is 6.20. Did any of our fictional characters have a greater annual dose?

   _____

4. The international standard for the annual radiation dose of those whose jobs bring them into regular contact with radioactive materials is 50 mSv. Assuming that any less than this amount of radiation is not a cause of concern, should any of these five people be concerned about their radiation exposure? Explain.

   _____

   _____

5. Of the sources for the largest dose of radiation for each person in the list in Table 1, which source delivered the largest dose?

   _____

6. What was the most common source for the largest dose of radiation?

   _____

7. Having a smoke detector in your house increases your exposure to radiation. Why do most people choose to install them in their homes?

   _____

   _____

8. According to the online radiation dose calculator, which results in a higher dose of radiation: living near a nuclear power plant or living near a coal-fired power plant?

   _____

9. Would moving away from a power plant significantly affect a person's annual radiation dose? Explain.

   _____

   _____

   _____

10. On the basis of your data, what are three things that someone could do to limit his radiation exposure?

   _____
   _____
   _____

> **GOING FURTHER**

Radon is a radioactive gas that is produced by the decay of uranium. It can collect in houses, especially in basements.

Happily, radon tests are commonly available, and there are structural modifications that can be made to make a house less prone to radon buildup.

11. What kind of health effects does radon exposure have?

    _____
    _____
    _____

12. How can a person know whether they have radon collecting in their home?

    _____

13. If a person discovers that his house has high radon levels, what can he do to fix the problem?

    _____
    _____

14. Is your county in a high risk area for radon buildup?

    _____

15. Why would a Christian be concerned with the amount of radon in houses?

    _____
    _____
    _____

Table 1

PERSON	ESTIMATED ANNUAL RADIATION DOSE (mSv)	LARGEST SOURCE OF RADIATION	AMOUNT OF SINGLE LARGEST DOSE (mSv)
Abner Brown			
Nemo Waterman			
Rebecca Batisaac			
Sidney Jug			
Artemis Vernon			

CHAPTER 9: SOLUTIONS

LAB 9A

ALL MIXED UP

Inquiring into Separating Mixtures

Historically, gold has been one of the most prized elements known. People have crossed oceans and continents hoping to strike it rich. Few ever have, but their ventures show the value that people have set on this lustrous yellow metal. Even today, gold can cost over $40 a gram.

Happily for gold miners of the past and present, gold does not typically react with other elements to form compounds. But it *is* typically found in a mixture with rock called *gold ore*. To be of any value, the gold must be separated from the mixture.

Separating gold from its ore is just one example of separating a mixture, a task scientists face very day. In this lab activity, you will plan and execute a process for separating a mixture of sand, iron filings, and sodium chloride.

QUESTIONS

How can I separate the components of a mixture?

- How can the properties of matter be used to separate a mixture?
- How do I know how well I separated the mixture?

Equipment

sand, iron filings, and sodium chloride mixture
goggles

PROCEDURE

Planning/Writing Scientific Questions

A Write a list of the physical properties of matter.

B Now list the physical properties of each of the three components of the mixture that you are considering.

C Write questions that will help guide your process of separating mixtures.

Designing Scientific Investigations

D Write procedures that will allow you to separate each of the components from each other. Think through all the steps needed for you to be able to mass the separated components when done. Your procedures must allow you to compare the combined mass of the recovered components with the mass of your original mixture sample.

E Have your teacher approve your procedures.

Conducting Scientific Investigations

F Obtain a sample of the mixture from your teacher.

G Follow your procedures to separate the components of the mixture.

Evaluating the Process

Percent composition is a comparison of the mass of a component to the mass of the whole.

$$\%_{composition} = \left(\frac{mass_{component}}{mass_{mixture}}\right) 100\%$$

1. Do you think that you should use the mass of the original mixture or the sum of the recovered masses as the $mass_{mixture}$ in the percent composition calculations? Explain.

   _____

   _____

H Use percent composition to assess your procedures and laboratory technique. Your teacher will be able to tell you the percent composition of the original sample.

I Use percent recovered to assess your procedures and laboratory technique.

$$\%_{recovered} = \left(\frac{mass_{recovered}}{mass_{original}}\right) 100\%$$

Name _____

Scientific Argumentation

J State a claim about how well your procedures and technique accomplished the separation of the mixture. Support your claim with evidence.

2. How could you change your procedures and technique to improve the results?

   _____
   _____
   _____

3. Evaluate your procedures using the data you collected.

   _____
   _____
   _____
   _____
   _____

GOING FURTHER

Typically, the liquids in a mixture will have different boiling points. One common process, called *fractional distillation*, takes advantage of this fact to separate the liquids in the mixture. In this process, the mixture is heated to the temperature at which the liquid with the lowest boiling temperature vaporizes. The vapor is then removed and condensed into a liquid. The process is repeated until all the liquids have been separated. Crude oil is commonly separated into separate liquids by fractional distillation.

4. Crude oil contains a variety of substances, including components of gasoline, motor oil, tar, and kerosene. Why would it need to be separated?

   _____
   _____
   _____

5. Why is fractional distillation important for resource management?

   _____
   _____
   _____

CHAPTER 9: SOLUTIONS

LAB 9B

THAT'S COLD!

Investigating Freezing Point Depression

You have probably seen your parents add antifreeze to the family car. Is this to keep the engine from getting too cold? Yes, but it also keeps the engine from getting too hot. It's a type of coolant.

Water makes a pretty good engine coolant too. But water has one problem. When it gets cold, it freezes. Solid ice does not make a good coolant. Even worse, water expands as it freezes, so it can burst parts of the engine, resulting in costly repairs. Antifreeze is a solution made with water that freezes at a much lower temperature than pure water. So most people outside the Arctic Circle don't have to worry about their coolant freezing overnight.

In this activity, you will investigate the freezing points of several solutions of equal concentrations.

QUESTIONS

How can I cool a solution below its solvent's normal freezing point?

- What is freezing point depression?
- Do different solutes affect freezing point depression differently?

Equipment
Erlenmeyer flasks, 250 mL (5)
graduated cylinder, 100 mL
ice bath
thermometer
wax pencil
distilled water
sugar ($C_{12}H_{22}O_{11}$) solution
sodium chloride (NaCl) solution
potassium chloride (KCl) solution
magnesium chloride ($MgCl_2$)
  solution
goggles

PROCEDURE

A Use the wax pencil to label the five Erlenmeyer flasks 1, 2, 3, 4, and 5.

B Use the graduated cylinder to fill each Erlenmeyer flask with 100 mL of the solutions listed in Table 1.

C Set the five flasks in the ice bath for twenty minutes.

1. What is the purpose of the flask of water?

2. What do you notice about the formulas of the compounds that are in the solutions?

3. The ice bath may have a temperature of −7.4 °C. Do you think that the solutions in any of the flasks will become this cold? Explain.

4. How is the ice bath able to reach a temperature as low as −7.4 °C without freezing?

5. Hypothesize about the final temperature of each of the flasks.

6. Why do you leave the flasks in the ice bath for twenty minutes?

D Measure the temperature of each flask, wiping off the thermometer between each measurement. Record the temperatures in Table 1.

7. Did any of the flasks have a different temperature than the others? Explain.

8. Did any flasks have the same temperatures? Explain.

9. Hypothesize an explanation of your observations from Questions 7 and 8.

   _____
   _____
   _____

10. Why did the ice bath get so cold?

    _____
    _____

GOING FURTHER

Freezing point depression is also used in making homemade ice cream. The milk, eggs, and sugar mixture that will form the ice cream is placed in a round metal container that is spun by hand or by a motor. The metal container is placed in a large bucket that is filled with ice and salt.

11. What is the purpose of the ice in the bucket?

    _____
    _____

12. Why would salt be added to the ice in the bucket?

    _____
    _____

13. Would it be possible to make ice cream without adding salt to the ice? Explain.

    _____
    _____

14. Read 1 Timothy 4:4. What does this passage teach us about making a nonnecessity like ice cream?

    _____
    _____

Table 1

FLASK	CHEMICAL	TEMPERATURE (°C)
1	distilled water	
2	$C_{12}H_{22}O_{11}$ solution	
3	NaCl solution	
4	KCl solution	
5	$MgCl_2$ solution	

CHAPTER 10: ACIDS, BASES, AND SALTS

LAB 10A
pH pHun

Determining pH

How can you tell whether a lemon is good enough to make lemonade? Acidic lemons make excellent juice, but lemons with low acidity don't. Growers in the 1930s tried to measure lemon acidity with various tests but couldn't get reliable numbers. They took the problem to chemistry professor Arnold Beckman to see whether he had any ideas.

With a strong background in chemistry and electronics, Beckman was well-qualified to solve the problem. He assembled a compact instrument that could measure acidity just by dipping a probe in lemon juice and reading the result from a meter. He called his device an "acidimeter." Later on, the name changed to pH meter, the name that we still use today.

Today, pH meters are commonly used instruments. In this lab activity, you will use one to explore the pH of some everyday substances. In doing so, you'll learn more about acids and bases. Perhaps you will also have a better appreciation for the conveniences that Beckman's work made possible.

QUESTIONS

Do pH meters and pH paper provide similar results?

- How can I measure the pH of a solution?
- What are the pH values of common solutions?
- How can knowing the pH of a solution be useful?

Equipment

Labdisc Gensci
pH electrode
plastic mini cups, (10)
beaker, 500 mL
wash bottle with distilled water
distilled water
ammonia
white vinegar
cola soft drink
lemon-lime soft drink
coffee
milk
calcium carbonate solution
sodium hydroxide, 0.1 M
hydrochloric acid, 0.1 M
universal pH indicator paper
goggles
laboratory apron
nitrile gloves

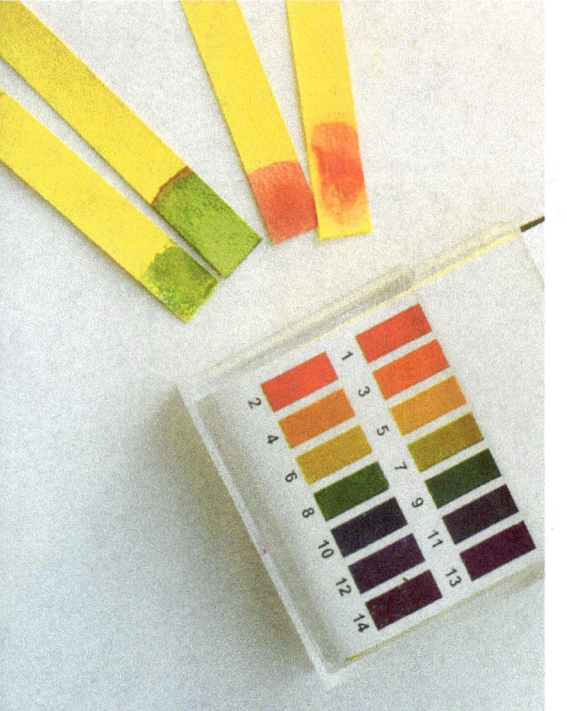

PROCEDURE

Universal pH Indicators

One of the easiest ways to evaluate pH is with a tool called an *indicator*. An indicator is a chemical that changes color in response to different pH values. The simplest pH indicator is litmus paper, which comes in two varieties, red and blue. Litmus paper's usefulness is limited, though, since it doesn't change color in the presence of mild acids or bases. It also doesn't give any suggestion of what the actual pH value of a solution is.

A more sophisticated test involves a special paper that contains several different indicator chemicals. The chemicals in the paper give a wider range of colors than litmus paper does, so the paper can measure pH values from 1 to 14. For that reason, it is called a *universal indicator*. To determine a substance's pH with a universal indicator, you must compare the paper to a color reference card (see bottom left).

A Obtain samples of each liquid to be tested (see Table 1).

B Tear off a small strip of universal indicator paper and briefly dip the end in the first liquid. After fifteen seconds, compare the color to the reference card that comes with the paper. If possible, use natural daylight illumination. Determine the pH to the nearest unit and record it in Table 1. If you think that the color falls in between two reference colors, express the pH as a range between those two values (e.g., 10–11).

C Repeat Step B for each liquid.

1. Can you use universal indicator paper to determine whether a substance is neutral? Explain.
   _____
   _____

2. What is a significant limitation of universal indicator paper?
   _____
   _____
   _____

3. In which type of situation would universal indicator paper be a useful tool?
   _____
   _____

4. Why would universal indicator paper *not* be a good choice when you're trying to determine the pH of a dark-brown, thick liquid?
   _____
   _____

5. Would universal indicator paper be a good way to measure the pH of household bleach, which has a pH between 12 and 13? Explain. (*Hint*: Think about the purpose of bleach.)

_____

_____

pH Meters

The most powerful and precise tool for measuring pH is the pH meter. It uses an electronic sensor to measure the quantity of hydrogen ions in a solution. While modern pH meters are smaller, cheaper, and more sophisticated than Beckman's original model, they work in much the same way. A good pH meter can measure pH to the nearest 0.01 unit or better.

D   Connect the pH sensor cable to the coaxial outlet on the Labdisc and turn it on. Click the pH button. Unscrew and remove the bottle of storage solution from the bottom of the probe. Slide the cap off the probe and set it and the storage solution bottle aside.

E   Squirt distilled water from the wash bottle onto the probe tip to rinse off the storage solution. Be sure to rinse both the glass bulb and the plastic shell surrounding it. Collect the waste water in the beaker or perform the rinse over a sink. Carefully and gently flick the probe to shake off excess water. The tip is fragile, so be very careful not to accidentally hit anything!

F   Dip the probe tip in the first liquid. Watch the pH display on the Labdisc screen. It should settle to a fairly stable value within about 10 seconds. Record this value to the nearest 0.01 unit in Table 1.

G   Repeat Steps E and F for each liquid. Always rinse the probe tip between each measurement using the technique outlined in Step E.

H   After you've tested the final liquid, give the probe an especially good rinse and flick off the water. Slide the storage bottle cap back onto the probe and screw the bottle back on. Be sure that the probe tip is near the bottom of the storage container and is totally covered by the storage solution.

I   Wash your hands well when you're finished cleaning up.

ANALYSIS

J   On the basis of your observations with the universal indicator paper and the pH meter, decide whether each liquid is acidic, basic, or neutral. Record your conclusions in Table 1.

CONCLUSIONS

6. What is a major advantage that the pH meter has over color-based indicators?

   _____
   _____
   _____

7. What is a disadvantage of the pH meter compared with color-based indicators?

   _____
   _____
   _____

8. In which kinds of situations would the pH meter be "overkill" compared with simpler methods?

   _____
   _____
   _____

GOING FURTHER

9. Swimming pools should have a pH that is slightly basic, typically greater than 7 but less than 8. Would a universal pH indicator be adequate for a pass-fail test of a pool's water? Explain.

   _____

   _____

10. When canning acidic foods, it's crucial that the pH always be less than 4.6, or dangerous bacteria can grow in the food. Would a color-based indicator be sufficient, or would a pH meter be a better choice? Explain.

    _____

    _____

    _____

    _____

11. On the basis of its pH, why do you think calcium carbonate makes a good antacid?

    _____

    _____

    _____

Table 1

SUBSTANCE	UNIVERSAL INDICATOR	pH METER	CONCLUSIONS
Distilled Water			
Ammonia			
White Vinegar			
Cola Soda			
Lemon-Lime Soda			
Coffee			
Milk			
Calcium Carbonate			
Sodium Hydroxide			
Hydrochloric Acid			

CHAPTER 10: ACIDS, BASES, AND SALTS

LAB 10B
FEELING THE BURN

Comparing the Concentrations of Basic Solutions

Have you ever experienced the consequences of one too many slices of pepperoni pizza? The discomfort of heartburn that often follows eating spicy foods is caused by excess stomach acid. After a pizza binge, a person might turn to an antacid for help. The ingredients in antacids neutralize stomach acid. The active ingredients in some antacids are *metal hydroxides*, ionic compounds that produce free hydroxide ions in aqueous solutions.

Is one metal hydroxide as useful as another for neutralizing acidic solutions such as stomach acid? In this lab activity, you'll use a technique called *titration* to determine how much of a metal hydroxide solution is needed to neutralize a solution of hydrochloric acid. On the basis of that information, you will then predict how much of a different metal hydroxide solution will be needed to perform the same task.

QUESTIONS

Do all basic solutions neutralize acid equally well?

- How can I tell how concentrated a solution is?
- What are key factors in determining how effective a basic solution is at neutralizing an acid?

Equipment

Labdisc Gensci
pH electrode
ring stand
magnetic stirrer
beaker, 150 mL
burette clamp
burette, 50 mL
graduated cylinder, 100 mL
distilled water
hydrochloric acid (HCl), 0.1 M
phenolphthalein solution, 0.02%
sodium hydroxide (NaOH, unknown concentration)
barium hydroxide (Ba(OH)$_2$), 0.1 M
paper towels
goggles
laboratory apron
nitrile gloves

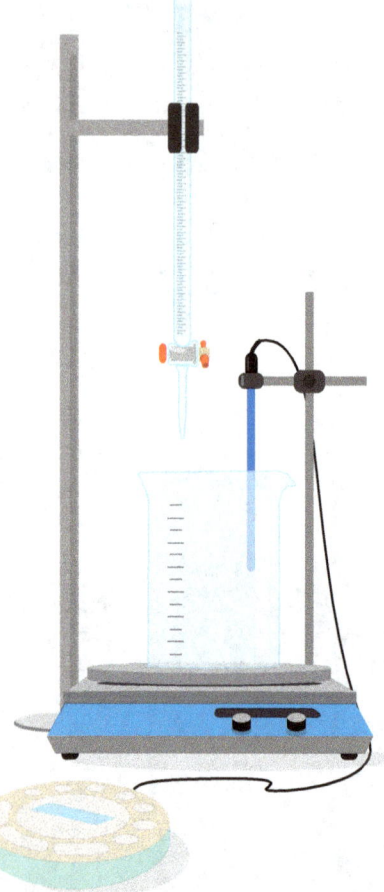

PROCEDURE

Determining the Concentration of a Solution

As you read in your textbook, antacids neutralize stomach acids, producing salt and water during the reaction. Stomach acid is a strong solution of hydrochloric acid (HCl). Most antacids contain mineral salts that produce hydroxide ions (OH$^-$) in solution. Though not used in antacids, sodium hydroxide (NaOH) is another strong producer of hydroxide ions. Hydrochloric acid and sodium hydroxide react as shown in the following equation.

$$HCl + NaOH \rightarrow NaCl + H_2O$$

1. Is this neutralization reaction balanced? Explain.

   _____

   _____

2. On the basis of your answer to Question 1, how many moles of sodium hydroxide do you think are needed to neutralize one mole of hydrochloric acid?

   _____

   _____

3. On the basis of your answer to Question 2, write a hypothesis about how many mL of sodium hydroxide solution will be needed to neutralize 40 mL of 0.1 M hydrochloric acid solution. Explain the reasoning for your hypothesis.

   _____

   _____

A Arrange the Labdisc, ring stand, magnetic stirrer, and beaker as shown on the left. Place the magnetic stir bar in the beaker.

B Connect the pH sensor cable to the coaxial outlet on the Labdisc and turn it on. Click the pH button. Unscrew and remove the bottle of storage solution from the bottom of the probe. Slide the cap off the probe and set it and the storage solution bottle aside.

C Squirt distilled water from the wash bottle onto the probe tip to rinse off the storage solution. Be sure to rinse both the glass bulb and the plastic shell surrounding it. Collect the waste water in a container or perform the rinse over a sink. Carefully and gently "flick" the probe to shake off excess water. The tip is fragile, so be very careful not to accidentally hit anything!

D Fill the beaker with 40 mL of hydrochloric acid and add five drops of phenolphthalein. Clamp the pH probe so that the tip is immersed in the acid solution but will not be struck by the stir bar nor touch the sides of the beaker.

E Fill the burette with 50 mL of sodium hydroxide solution. This is your *titrant*, the solution that is being added to the system.

F Turn on the magnetic stirrer and adjust the setting for a gentle stir. You are now ready to titrate!

G Add sodium hydroxide to the beaker of hydrochloric acid in 1 mL increments. Observe the pH readout on the Labdisc as you proceed with the titration.

H When the pH readout on the Labdisc reaches 6.0, begin adding titrant dropwise instead of in milliliter increments. Add drops until the solution in the beaker reaches a pH of 7.0.

I Follow your teacher's instructions for disposing of the used chemicals.

4. What is the total amount of titrant that you added to the beaker of hydrochloric acid?

ANALYSIS

5. Did your answer to Question 4 support your hypothesis in Question 3? Explain.

6. What does the final pH reading of 7.0 on the Labdisc tell you about the reaction in the beaker?

7. Is it absolutely necessary for this activity that a pH probe be used? What other method might be useful?

GOING FURTHER

There are of course other basic solutions that could be used to neutralize a sample of hydrochloric acid. Use what you learned in the previous section to consider how this might work.

8. On the basis of your results in the previous section, what quantity of 0.1 M barium hydroxide do you think will be required to neutralize 40 mL of 0.1 M hydrochloric acid? Write your prediction in the form of a hypothesis.

9. Write out a balanced chemical equation for the neutralization of 0.1 M hydrochloric acid with 0.1 M barium hydroxide. (*Hint*: Make sure that you use the proper chemical formula for barium hydroxide!)

_____

10. According to your answer to Question 9, how many moles of barium hydroxide are required to neutralize one mole of hydrochloric acid?

_____

_____

11. Is your answer to Question 10 consistent with your hypothesis in Question 8?

_____

12. Is barium hydroxide better or worse at neutralizing hydrochloric acid than sodium hydroxide? Defend your claim using your answer to Question 10.

_____

_____

_____

13. On the basis of what you have learned from this activity, why can you *not* predict the amount of a base needed to neutralize a particular amount of an acid on the basis of the molarities of the two solutions alone?

_____

_____

_____

14. Using what you have learned from this activity, which do you think would make a more effective active ingredient in an antacid, sodium hydroxide or barium hydroxide? Explain.

_____

_____

_____

15. Do an internet search using the keywords "antacid active ingredients." Is barium hydroxide used in antacids? Why or why not?

_____

_____

CHAPTER 11: KINEMATICS

LAB 11A

WAY TO GO

Inquiring into Distance and Displacement

Global navigation around the earth is possible because scientists have created a grid system on the earth using latitude and longitude. The *Global Positioning System* (GPS) uses this grid system as a frame of reference, like a coordinate plane. Modern airliners travel directly between any two points on the globe by using this coordinate system. NASA engineers use a similar coordinate system to direct robots to explore the surface of Mars.

In this lab activity, you will determine how to direct another lab group to move around your school. You will have to decide how to provide those instructions and what your frame of reference will be.

QUESTIONS

How can we navigate from place to place?

- What information is needed to write directions to travel between two given points?
- How do I create a scale map?
- What is the criteria for evaluating directions?

PROCEDURE

Planning/Writing Scientific Questions

A Get starting and ending points from your teacher. Write them both below. Record the starting point also in Table 1.

  Starting point: _____

  Ending point: _____

B With your lab group, determine the tools that you will need to write directions from your starting point to your ending point.

  1. How are you going to describe turn directions for your instructions?

  _____

  _____

Equipment

computer or graph paper
metric ruler
protractor

Way to Go 107

2. How are you going to describe distances for your instructions?

   _____
   _____

Executing Your Plan

C Go to your starting point and enter information in Table 1 about the way to face when you start your instructions. For example, if you are starting at the main office door you might say, "Stand facing the main office." *This is the only time that you can refer to the building.*

D Record your directions in Table 1. Remember not to reference the building in any way in your directions.

Developing Models

Modern GPS systems not only provide instructions for traveling from place to place, they also include a moving map to help visually orient a traveler to his path. You will do the same thing by drawing a scale map of your instructions.

E Determine the scale needed to complete your map on a single piece of computer paper.

3. What is the scale for your map?

   _____

F Create the scale model of your directions.

4. What was the total distance traveled?

   _____

5. What is the displacement from the starting point to the destination on your map?

   _____

6. What is the displacement from the starting point to the destination in the school?

   _____

G Turn in your directions and map to your teacher.

Following Directions

H Get a set of directions from your teacher.

I Beginning at the assigned starting point, follow the directions. When you arrive, record your position below.

   Destination: _____

7. What challenges did you face in following the directions?

   _____
   _____
   _____

Scientific Argumentation

J Ask your teacher for the expected destination. Record it below.

   Expected Destination: _____

8. Write a statement evaluating the directions that you were to follow. Remember that scientific statements should include your claim *and* evidence for the claim. For example, "The directions we were given did a good job directing us to our destination. We ended up arriving at a location that was 1.2 m from the intended destination."

   _____
   _____
   _____

9. How can navigation systems and navigation skills be useful in fulfilling the Creation Mandate?

   _____
   _____
   _____

Table 1

STARTING POINT:		FACING:	
LEG	DIRECTION		DISTANCE
1			
2			
3			
4			
5			
6			
7			
8			
9			
10			

CHAPTER 11: KINEMATICS

LAB 11B
SLOW AND STEADY

Investigating Uniform Motion

Your textbook used Aesop's fable "The Tortoise and the Hare" to discuss velocity versus time graphs. In the fable, the speedy hare raced out to a huge lead. His lead was so great that he figured that he would have time to take a nap and still win the race. While the hare was napping, the tortoise won the race by moving at his slow, steady speed for the entire race. While there is debate about the lesson of the fable, physicists would say that the tortoise demonstrated uniform, unchanging motion.

What are some things that move at a constant speed? A car traveling down the highway with its cruise control on? An airplane flying on autopilot? But do these machines truly maintain a constant speed? In this lab activity, you will investigate uniform motion. You will record data for a ball rolling at a constant speed across the floor. You will even see whether one of your classmates can walk at a constant rate.

QUESTIONS

Can we achieve uniform motion?

- What is uniform motion?
- What affects uniform motion?
- Why is uniform motion so difficult to produce?

Slow and Steady

PROCEDURE

Set Up the Labdisc

A Set up the Labdisc to collect distance data. For help with setting up the Labdisc, refer to Appendix F.

B Set the **Sampling rate** to **25/sec**.

C Set the **Number of samples** to **100**.

1. If the Labdisc is taking 25 samples every second and you are collecting 100 samples, for how many seconds will the Labdisc collect data?
   _____

Collecting Data—Ball

D On the bottom of the Labdisc, open the cover for the distance sensor.

E Hold the Labdisc at the end of the table with the distance sensor cover lying flat on the tabletop and the distance sensor pointing along the table surface.

F Begin data collection and then have a member of your lab group roll the ball directly away from the sensor. The ball must start at least 0.2 m from the sensor.

G Close the cover for the distance sensor.

2. What do you expect a position versus time graph of an object moving with constant, uniform velocity to look like?
   _____

3. Do you expect that the ball moved uniformly? Explain.
   _____
   _____
   _____

Analyzing Data

H Connect the Labdisc to your computer and download the last data set.

In most cases odd-looking data will appear at the start and end of the graph, but you should notice a smooth curve in the middle of the graph.

4. What do you think caused the odd-looking data at each end of the graph?
   _____
   _____

I Mark each end of the smooth curve.

Equipment
Labdisc Gensci
computer
ball

J Calculate the displacement traveled by subtracting the position at the left end of the smooth curve from the position at the right end. Calculate the time interval by subtracting the time at the left mark from the time at the right mark. Record both in Table 1.

5. Does the smooth section of the graph appear to be showing uniform motion?

   _____

6. If the graph doesn't appear linear, what do you think is causing the motion to change?

   _____
   _____
   _____

K Put a linear curve of best fit on the graph. Read the slope (the coefficient of the linear term), which represents the average velocity of the ball. Record the average velocity in Table 1.

L Use the slope tool to determine the slope at the left end of the smooth section, which represents the ball's initial velocity. Record the initial velocity in Table 1.

M Use the slope tool to determine the slope at the right end of the smooth section—the ball's final velocity. Record this value in Table 1.

7. Do the three velocity values make sense? Explain.

   _____
   _____

Collecting Data—Human

Do you think that you can establish uniform motion by walking at a constant rate?

N Disconnect the Labdisc from the computer and reopen the cover for the distance sensor.

O Set the display on the Labdisc to display distance data.

P Stand approximately 10 m from—and facing—a wall.

Q Begin data collection and then walk toward the wall, trying to maintain a constant rate.

R Reconnect the Labdisc to your computer and download the last data set so that you can view the data.

8. Were you able to establish uniform motion while walking? Explain.

   _____

GOING FURTHER

S   Use the graphing area on the next page to make a velocity versus time graph using the ball-initial velocity, the ball-ending velocity, and the time interval.

T   Use the formula for the area of a trapezoid,

$$A = \tfrac{1}{2}(b_1 + b_2)h,$$

where b_1 is the initial velocity, b_2 is the final velocity, and h is the time. The area under the curve is the displacement. Record this value in Table 1.

9. How does this graph match with what you saw on the position versus time graph?
   _____
   _____

10. How does your calculated displacement compare with the measured displacement?
    _____
    _____

11. Why do you think your walking graph appeared as it did?
    _____
    _____
    _____

12. Suggest a way that we could improve on the ball-rolling part of the lab activity to get the motion to be more linear.
    _____
    _____

Table 1

Measured Displacement (m)	
Time Interval	
Ball Average Velocity (m/s)	
Ball Initial Velocity (m/s)	
Ball Final Velocity (m/s)	
Calculated Displacement (m)	

Name

Slow and Steady

CHAPTER 11: KINEMATICS

LAB 11C

THE GRAVITY OF THE SITUATION

Investigating Free Fall

A skydiver steps through the door of the aircraft. He immediately begins to accelerate downward at 9.81 m/s². In the absence of air resistance, he would continue accelerating until reaching the ground. Happily, there is air resistance, which, upon the opening of the parachute, slows his descent so that he can make a perfect landing.

While many of us have never jumped out of an airplane, we have all experienced *acceleration due to gravity*. If you have ever tripped and fallen, you have moved with acceleration due to gravity. When we trip, gravity leads to a bad outcome. But gravity is vital for us to live. It is the force of gravity that holds the atmosphere around the earth. Without gravity there would be no air to breathe.

In this lab activity, you are going to investigate acceleration due to gravity. You will examine a position versus time graph for accelerated motion. You will also experimentally calculate acceleration due to gravity.

QUESTIONS

Can we determine the acceleration due to gravity?

- What does accelerated motion look like on a position versus time graph?
- Where can we see the acceleration due to gravity on a position versus time graph?
- How close to free-fall acceleration can we get?

Equipment

Labdisc Gensci
computer
ball

PROCEDURE

Set Up the Labdisc

A Set up the Labdisc for distance data collection.

B Set the **Sampling rate** to **25/sec**.

C Set the **Number of samples** to **100**.

Collecting Data

D On the bottom of the Labdisc, open the cover for the distance sensor.

E Hold the Labdisc upright with the distance sensor pointing straight down toward the ground.

F Have a member of your lab group hold a ball at least 0.20 m below the distance sensor. Begin data collection and then have him drop the ball.

G Close the cover for the distance sensor.

1. What do you expect the position versus time graph of an accelerating object to look like?

   _____

Analyzing Data

H Connect the Labdisc to your computer and download the last data set.

I Place your cursor over the **Distance** on the *y*-axis and right click. Click **Reverse**. You may need to adjust your maximum and minimum values.

In most cases, odd-looking data will appear at the start and end of the graph, but then you should notice a smooth, downward curve followed by a series of smooth, inverted u-shaped curves in the middle of the graph.

2. Why do you think the inverted u-shaped sections of the graph change as they do?

   _____

   _____

J Mark the ends of the smooth curve.

K Add a quadratic curve of best fit on the graph. Read the coefficient of the quadratic term and record this value in Table 1.

L Right click on each of the two textboxes for the end points to delete the textboxes.

M Repeat Steps J–L for three of the inverted u-shaped sections of your graph.

3. Compare the four coefficients. Does your sample seem to be giving you consistent data?

    _____

4. What do you think are some reasons that the data varied as it did?

    _____

    _____

N Calculate the average of the coefficients and enter that value in Table 1.

O Multiply the average of the coefficients by 2 to calculate your measured value for acceleration due to gravity and enter this value in Table 1.

P The accepted value for acceleration due to gravity is 9.81 m/s². Calculate the percent error and enter this value in Table 1.

$$\% \ error = \frac{(gravity_{measured} - gravity_{accepted})}{gravity_{accepted}}(100\%)$$

5. Was your calculation of acceleration due to gravity accurate?

    _____

GOING FURTHER

6. The graphs of motion that we make are models of the real world. It is important to understand the connections between the model and the real world. Look at the graph of the motion again. What do the sharp angles at the ends of the smooth curves represent?

    _____

    _____

    _____

    _____

7. What does the top of the smooth curve represent?

    _____

    _____

    _____

    _____

8. Why do you think that the height at the top of each smooth curve gets lower each time?

   _____
   _____
   _____

9. What does the slope of this graph represent?

   _____
   _____
   _____

10. Can you think of two things that would be different if God hadn't created the force of gravity?

    _____
    _____
    _____

Table 1	
Coefficient 1	
Coefficient 2	
Coefficient 3	
Coefficient 4	
Average Coefficient	
Acceleration Due to Gravity (m/s^2)	
Percent Error (%)	

CHAPTER 12: DYNAMICS

LAB 12 A
LAB HEARD ROUND THE WORLD

Investigating the Second Law of Motion

Any competitive shooter knows to pull her rifle tight into her shoulder before she fires. If she doesn't, the rifle will recoil rapidly into her shoulder. Think about this as you perform this lab activity, and see whether you can recognize how the second law of motion explains this phenomenon.

As you work through this activity, you will experiment with the effects of mass and force on acceleration.

QUESTIONS

How are force, mass, and acceleration related?

- How does a change in force affect acceleration?
- How does a change in mass affect acceleration?

PROCEDURE

Collecting Data

A Set up the Labdisc to collect data from only the distance sensor. (Refer to Appendix F for questions on using the Labdisc.)

B Set the **Sampling rate** to **25/second**.

C Set the **Number of samples** to **100**.

D Make sure that the Labdisc is displaying *Distance, m*.

Equipment
Labdisc Gensci
dynamics cart
computer
pulley board
string
mass set
meter stick

Lab Heard Round the World 121

E Set up the pulley board according to the image below. Tie one end of the string to the dynamics cart and the other end to the mass hanger with a 150 g mass. Run the string over the pulley.

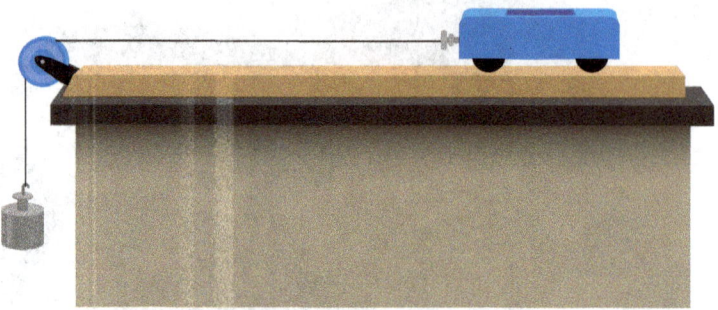

F Place a 100 g mass in the dynamics cart and secure it with tape if necessary. Mass the dynamics cart and record the value in Table 1.

G On the bottom of the Labdisc, open the cover for the distance sensor.

H Hold the Labdisc at the end of the pulley board with the distance sensor cover lying flat on the board and the distance sensor pointing along the board's surface.

I Pull the dynamics cart back to 0.3 m from the end of the pulley board.

J Begin collecting data and then release the dynamics cart.

K Close the cover for the distance sensor.

Force and Acceleration

L Start the Globilab software and download the last data set.

In most cases, you should notice a smooth curve in the middle of the graph. There may be odd-looking data at the very end of the curve.

1. What do you think caused the odd-looking data at the end of the graph?

   _____

   _____

M Place markers at the left and right ends of the smooth curve.

N Overlay a quadratic curve of best fit on the graph.

Note that the formula for the quadratic curve is in the form of

$$At^2 + Bt + C.$$

The formula below is a form of the distance formula.

$$\mathbf{d} = \frac{1}{2}\mathbf{a}t^2 + \mathbf{v}_0 t + \mathbf{d}_0$$

Note that $\mathbf{d}$ is final displacement, $\mathbf{a}$ is acceleration, t is time, $\mathbf{v}_0$ is initial velocity, and $\mathbf{d}_0$ is initial displacement.

2. If the formula for the quadratic curve is $d = 5.0t^2 - 3.4t + 1.2$, what would the acceleration be?

O. Read the coefficient of the t^2 term, which represents the acceleration of the cart. Record the acceleration in Table 1.

3. Write a hypothesis about what will happen to the acceleration of the cart as the force applied to it decreases.

   _____
   _____

P. Repeat Steps E–O for the remaining masses listed in Table 1.

4. How does changing the mass on the mass hanger change the force applied to the dynamics cart?

   _____
   _____

5. According to your data, what effect did a change in the force applied by the mass hanger have on the cart's acceleration?

   _____
   _____

6. Does your data support your hypothesis? Use examples from your data to support your claim.

   _____
   _____

Mass and Acceleration

7. Write a hypothesis about what will happen to the acceleration of the cart if the mass of the cart is decreased.

   _____
   _____

Q. Place a 100 g mass in the dynamics cart. Mass the dynamics cart and record the value in Table 2.

R. Place a 100 g mass on the mass hanger.

S. Repeat Steps E–O, recording the cart's acceleration in Table 2.

T. Repeat Steps Q–S using 80 g, 60 g, 40 g, and 20 g masses.

8. According to your data, what effect did a change in the cart's mass have on its acceleration?

   _____

   _____

9. Does your data support your hypothesis? Use examples from your data to support your claim.

   _____

   _____

GOING FURTHER

When a rifle fires, the force of the expanding gases produced by the burning powder pushes the bullet forward. The same force pushes the rifle backward. This acceleration backward is called *recoil*.

10. Given the second law of motion formula, **F** = m**a**, which will the force of the expanding gases accelerate more, a 3 kg rifle or a 4 kg rifle? Explain.

    _____

    _____

11. On the basis of your answer to Question 10, how might a shooter holding his rifle tight to his shoulder affect the rifle's recoil?

    _____

    _____

12. Why would a shooter holding his rifle tight to his shoulder reduce the chance of injury to the shooter?

    _____

    _____

Table 1

CART MASS (g)	HANGING MASS (kg)	FORCE APPLIED (N)	CART ACCELERATION (m/s²)
	0.150	1.470	
	0.130	1.274	
	0.110	1.078	
	0.90	0.882	
	0.70	0.686	

Table 2

HANGING MASS (kg)	FORCE APPLIED (N)	CART MASS (kg)	CART ACCELERATION (m/s²)
0.100	0.981		

CHAPTER 12: DYNAMICS

LAB 12B
ROUGH GOING

Investigating the Properties of Friction

Imagine a world without friction. A car owner wouldn't have to worry about putting oil in his car to prevent engine wear. But the car also wouldn't go anywhere because the tires wouldn't interact with the ground. And the engine wouldn't be able to turn the wheels in the first place. Even if the car owner did manage to get the car going by attaching a small rocket engine to it, he would have a very hard time stopping since brakes work by friction too.

But friction isn't always a help. The engines in most cars convert only about 25% of the chemical energy in gasoline into usable mechanical energy. The other 75% is lost as thermal energy. And the force responsible for much of that wasted energy is friction.

So friction can be both beneficial and detrimental. In this lab activity, you will investigate how mass and surface area affect the friction acting on an object.

QUESTIONS

How much force does friction exert?

- What is friction?
- How do weight and surface area affect friction?

PROCEDURE

1. Are mass and surface area physical or chemical properties?

Friction and Weight

You will first explore the role that weight plays in friction. Recall that weight is the force of gravity acting on a mass.

Equipment
laboratory balance
spring scale
friction block
500 g masses (4)
metric ruler

Rough Going 125

2. Write a hypothesis that you think expresses the relationship between weight and friction.

   _____
   _____

   A Mass the friction block and record it in Table 1.

   B Use the spring scale's metal loop to hook it to the friction block's screw eye.

   C Set the friction block on a flat surface. Pull the spring scale and attached block steadily by its metal hook, keeping the spring scale parallel to the direction of motion.

   D Record the amount of force that you are using to pull the block in Table 1.

   E Repeat Steps A–D four times, adding 500 g more mass to the block each time.

   F Plot the data in Table 1 in the graphing area on page 128. Place the mass on the x-axis and the force on the y-axis. Draw a smooth curve for the data.

3. On the basis of your data in Table 1 and the graphing area, describe the relationship between friction and weight.

   _____
   _____

4. Does the data from this test support your hypothesis?

   _____
   _____

Friction and Surface Area

In this test, you will explore whether surface area plays any role in friction.

5. Write a hypothesis that expresses a relationship between surface area and friction.

   _____
   _____

   G Measure the length and width of the friction block's front (large) surface. Record these values in Table 2.

   H Calculate the front surface's area and record the result in Table 2.

   I Repeat Steps G and H for the friction block's side (small) surface. Record these values in Table 2.

J Lay the friction block on its front surface. Without adding any extra mass to the friction block, determine the force of friction using the same process that you used in the previous section. Record the value in Table 2.

K Repeat Step J with the friction block placed on its side surface.

6. On the basis of the data from Table 2, describe the relationship between friction and surface area.

   _____

   _____

7. Does the data from this test support your hypothesis?

   _____

   _____

GOING FURTHER

Air is a notable cause of friction. We don't normally notice air resistance when walking around, but any object that is traveling quickly must deal with air resistance.

8. When a meteor falls through the atmosphere, friction from air resistance heats up the meteor, resulting in the glow associated with meteors. Why is this a benefit?

   _____

   _____

9. How is God's wisdom displayed by air resistance on meteors?

   _____

   _____

Table 1

NUMBER OF MASSES ON BLOCK	MASS (g)	FORCE (N)
0		
1		
2		
3		
4		

Table 2

SURFACE	LENGTH (mm)	WIDTH (mm)	AREA (mm²)	FRICTION FORCE (N)
Front				
Side				

CHAPTER 13: *WORK AND MACHINES*

LAB 13A
A CLEAR ADVANTAGE

Investigating Pulleys

A crane can lift a 25-ton container off a dock and onto a ship preparing to travel across the ocean. A crane is used to do this work because it would take at least 350 strong men to move the container. But the crane is just a sophisticated combination of simple machines that humans have used for thousands of years.

In this lab activity, you will work with several pulley systems and observe the benefits and costs of using pulleys.

QUESTIONS

How does a pulley make work easier?

- How is a pulley constructed?
- What do we mean when we say that work is easier?
- How does adding extra pulleys affect mechanical advantage?
- What is the cost of using pulleys?

PROCEDURE

A Set up the single pulley, the string, a 1 kg mass, and the spring scale in the arrangement shown in the image at right. The length of the string should be about 4.5 times the height of the pulley.

B Use the spring scale to measure the input force needed to slowly raise the mass and record it in Table 1.

C Starting with the load on the floor and the string tight, lift the load off the floor by pulling the string at least 50 cm. Use the meter stick to measure the input distance and record it in Table 1.

Equipment

single pulley
1 kg mass
mass hanger
spring scale
meter stick
double pulleys (2)
string

A Clear Advantage

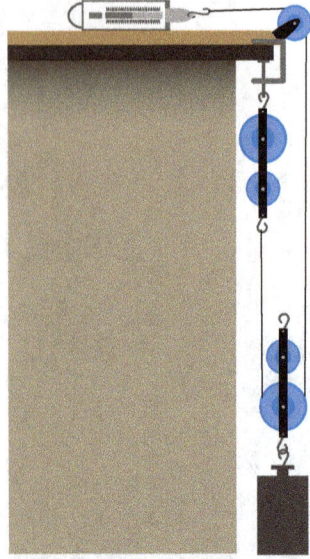

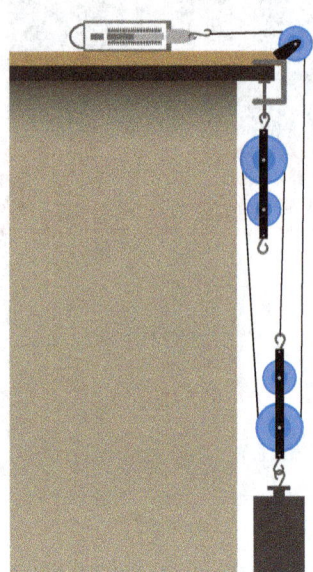

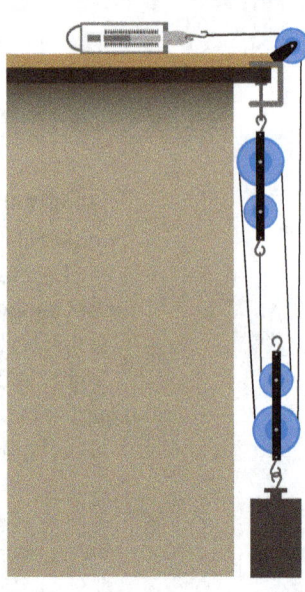

D Measure the height that the mass was raised when you pulled the string in Step C. Record this output distance in Table 1.

1. What did the single pulley do to the input force? Consider both magnitude and direction.

   _____
   _____

E Repeat Steps A–D for two, three, and four pulleys as shown in the images on the left.

2. What did the multiple pulleys do to the input force, compared with the output force?

   _____
   _____

3. On the basis of your answers to Questions 1 and 2, write a statement about how pulleys make work easier.

   _____
   _____

F Use the *AMA* formula below to calculate the actual mechanical advantage for each of your pulley systems and record these values in Table 1.

$$AMA = \frac{F_{out}}{F_{in}}$$

G Use the *IMA* formula below to calculate the ideal mechanical advantage for each of your pulley systems and record these values in Table 1.

$$IMA = \frac{d_{in}}{d_{out}}$$

H Calculate the *efficiency* of each pulley system by using the following formula.

$$efficiency = \left(\frac{AMA}{IMA}\right) 100\%$$

Record the *efficiency* of each pulley system in Table 1.

4. Is there a trend to efficiency as the number of pulleys increases?

   _____
   _____

5. What do you think causes this trend?

   _____
   _____

6. Given the trend that you noticed in Question 4, why would someone add an additional pulley to a system?

___

7. Could a system exist where the addition of another pulley would not make the work any easier? Explain.

___

8. How much work is put into the two-pulley system? (Use the formula $W = Fd$.)

9. How much work is used by the two-pulley system to lift the mass?

10. How much work was lost to friction in the two-pulley system?

11. Given that any use of a pulley system results in some work being lost due to friction, why would someone choose to use a pulley instead of lifting an object by hand?

___

GOING FURTHER

Pulleys are all around us, though we often don't see them because they are buried inside larger machines. But they are integral parts of conveyor belts, elevators, exercise machines, and many other systems.

12. Choose from a conveyor belt pulley, an elevator pulley, an exercise machine pulley or another pulley and do some research. Write a paragraph explaining how pulleys are used to accomplish the machine's purpose.

_____
_____
_____
_____
_____

Table 1

NUMBER OF PULLEYS	OUTPUT FORCE (N)	INPUT FORCE (N)	INPUT DISTANCE (cm)	OUTPUT DISTANCE (cm)	AMA	IMA	EFFICIENCY %
1	9.81						
2							
3							
4							

CHAPTER 13: WORK AND MACHINES

LAB 13B
RAMPING UP

Experimenting with Inclined Planes

The 1990 Americans with Disabilities Act (ADA) requires businesses and other public facilities to provide accommodations for people with disabilities so that they can access and use the resources. One of the most visible of these accommodations is ramps that allow people who cannot climb stairs easily to access buildings.

Some ramps are steeper than others. In this lab activity, you will examine how the steepness of a ramp affects its ability to improve a person's ability to perform work.

QUESTIONS

Why is a steep ramp harder to climb than a less steep one?

- How does an inclined plane make work easier?
- What is the cost of using an inclined plane?
- How can I evaluate the effectiveness of inclined planes?

PROCEDURE

1. Write a hypothesis regarding the effect that the slope of a ramp will have on its mechanical advantage.

   _____

   _____

   A Attach the 1 kg mass to the dynamics cart.

   B Use the laboratory balance to measure the mass of the dynamics cart and mass together. Multiply this mass by 9.81 m/s² to determine the weight of the cart and mass. This is the weight that the ramp is lifting—the output force. Record this value in Table 1.

Equipment

laboratory balance
1 kg mass
dynamics cart
spring scale
mass hanger
meter stick
ramp
books (5)
poster putty

Ramping Up   133

C. Zero the spring scale and attach the dynamics cart to the spring scale's hook.

2. Why is the length of the ramp in reality the input distance?
___
___

D. Use the meter stick to measure the input distance (ramp length) and record the value in Table 1.

E. Set up the ramp on a large book as shown in the figure below. Place a strip of poster putty along the bottom edge.

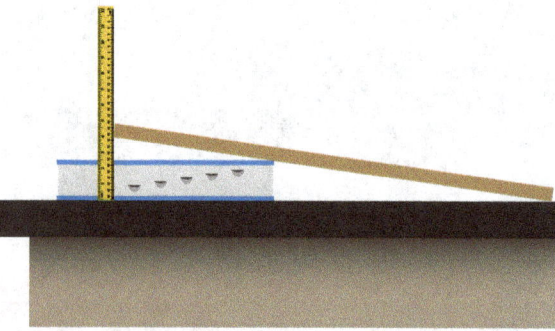

F. Use the meter stick to measure the height from the tabletop to the lower edge of the ramp—the output distance—as shown in the figure above. Record this value in Table 1.

3. Why should you measure only to the lower edge of the ramp?
___
___
___

4. Why is the measured ramp height in reality the output distance?
___
___

Now, collect some data by running some trials.

G. Place the dynamics cart at the bottom of the ramp and use the spring scale to pull the cart at a steady rate up the ramp. Record the force indicated on the spring scale in Table 2. Repeat twice more and average the three results. Then record the average input force in Table 1.

5. Why is it necessary to collect the data three times and then to average the results?
___
___

H. Repeat Steps E–G with 2, 3, 4, and 5 books under the ramp.

134   Lab 13B

I   Use the *AMA* formula below to calculate the actual mechanical advantage for each of the ramp positions and record these values in Table 1.

$$AMA = \frac{F_{out}}{F_{in}}$$

J   Use the *IMA* formula below to calculate the ideal mechanical advantage for each of the ramp positions and record these values in Table 1.

$$IMA = \frac{d_{in}}{d_{out}}$$

6. What trend do you notice in the mechanical advantage related to the ramp's steepness?

   _____
   _____

7. Did the data support your hypothesis?

   _____
   _____

8. Does this explain why a less steep ramp is easier to climb than a steeper one? Explain.

   _____
   _____

9. How would doubling both the height and length of a ramp affect the ramp's *AMA*?

   _____
   _____
   _____

10. How much work is put into the three-book ramp? (Use the formula $W = Fd$.)

11. How much work is done by three-book ramp to lift the mass?

K Calculate the efficiency of each ramp by using the formula below.

$$efficiency = \left(\frac{AMA}{IMA}\right)100\%$$

Record the efficiency of each ramp in Table 1.

12. What do the efficiency calculations show about the relationship between the ramp's *AMA* and its *IMA*?
    _____
    _____

13. What other fact could have told you the same thing?
    _____
    _____

14. Why was efficiency less than 100%?
    _____
    _____

15. Do you notice a trend in the efficiency data? Explain.
    _____
    _____

16. Since people using a ramp do more work than if they just lifted the object, why do people use them?
    _____
    _____

GOING FURTHER

Ramp steepness is an important consideration when planning accommodations for people with disabilities. One of the requirements established under the ADA is that an access ramp may not have a rise-run ratio greater than 1:12. In other words, a ramp that is 12 m long cannot have a height greater than 1 m.

17. Why is it necessary to set a maximum steepness under the ADA?
    _____
    _____

18. How is providing a ramp with a rise-run ratio less than 1:12 fulfilling the second great commandment?
    _____
    _____

Name _____

Table 1

RAMP POSITION	OUTPUT FORCE (N)	INPUT DISTANCE (m)	OUTPUT DISTANCE (m)	INPUT FORCE (N)	AMA	IMA	EFFICIENCY %
1							
2							
3							
4							
5							

Table 2

| RAMP POSITION | INPUT FORCE (N) | | |
	TRIAL 1	TRIAL 2	TRIAL 3
1			
2			
3			
4			
5			

CHAPTER 14: ENERGY

LAB 14A
HOLD YOUR HORSES

Investigating Work, Energy, and Power

In the early 1770s, James Watt had a marketing problem. Mine owners had been using inefficient Newcomen steam engines for over fifty years. Watt's new engines were much better, but he had no easy way to demonstrate just how much better they were. Unless customers could see the difference, they likely wouldn't buy one. Mine owners knew how much work a horse could do, so Watt compared a steam engine's power to a horse's power. He called this power measurement unit a *horsepower*. It's still used today.

In this lab activity, you will examine work, energy, and power. Some people think that they mean essentially the same thing, but each term has a distinct meaning. As you work through the activity, you'll learn what each term really means and when to use each one.

QUESTIONS

If I do a task in half the usual amount of time, aren't I doing twice as much work?

- What is the difference between work, energy, and power?
- How do I measure work, energy, and power?
- What is my "personal" horsepower?

PROCEDURE

Work and Energy

As you read in your textbook, when a force makes something move, work is being done. Energy makes work possible. Scientists sometimes descriptively define energy as the ability to do work. There are many different forms of energy, but they're all expressed in the same unit—the joule (J). Since energy does work, we express work in joules too.

When you climb a staircase, you use energy. Gravity pulls you down with a force proportional to your body's mass. Your muscles do work by opposing gravity's force and moving your body upward. It takes energy to do work, so your body supplies chemical energy from food to your muscles. Let's do some measurements to see how much energy you're using.

Equipment
staircase
metric ruler
bathroom scale
stopwatch

A Weigh yourself on the bathroom scale. The scale doesn't measure your mass. Instead, it measures your weight (w)—the force caused by gravity acting on your body's mass. If the scale reads in pounds, you will need to convert the indicated weight to newtons. Use the conversion factor below to do this.

$$\frac{4.6 \text{ N}}{1 \text{ lb}}$$

If your scale reports weight in kilograms, recall that an object's weight in newtons is equal to its mass—what your scale is reporting as "weight"—times its acceleration due to gravity (9.81 m/s² on Earth). Record your weight in Table 1.

B Next, you need to know the distance that you will move against gravity. Recall that the force of gravity is a downward vector quantity. The distance that we are concerned with is the vertical distance from the bottom of the staircase to the top. Using the ruler, measure the distance from the top of one step to the top of the next. Record it in Table 1.

C Count the total number of steps and record this value in Table 1. Finally, multiply this number by the height value that you obtained in Step B. Record the result in Table 1. This is the total vertical distance that you'll travel as you climb the steps.

D Use the following formula to calculate the work that you'll do climbing the stairs. Round your result to the nearest 10 J and record it in Table 1.

$$W = wd$$

1. What would happen to the amount of work that you'd have to perform if the staircase had twice as many steps? Explain.

   _____

   _____

2. Imagine yourself back in grade school. You weigh half your present weight. What would happen to the amount of work that you'd have to perform to climb the original staircase? Explain.

   _____

   _____

3. A popular brand of soda contains 140 calories of food (chemical) energy in a 12 oz can. That's equivalent to 585 800 J of energy! How many times could you climb the staircase before you used up the food energy in the can of soda? Round your answer to the nearest whole number.

   _____

   _____

4. On the basis of your answer to Question 3, explain how overconsumption of sugary drinks like soda can contribute to excess weight gain.

_____
_____
_____
_____
_____

POWER

Take a look at the equation for work and consider this question: If you climb the staircase slowly, will you use less energy than if you climb it quickly? Perhaps surprisingly, the answer is no. Time doesn't appear in the work equation. Whether you climb the staircase quickly or slowly, the amount of work that you do, and thus the amount of energy that you use, is exactly the same. Maybe that seems wrong. After all, you know from experience that you get tired sooner if you do something quickly. Walking is easier than running. So what's going on?

The answer rests on the rate at which work is done. If you climb the staircase quickly, your body must supply the required energy more rapidly than if you climb it slowly. When you do a lot of work in a short time, your body may have difficulty delivering the required energy fast enough as well as getting rid of the metabolic waste products. The result? You get tired.

You should recall from Chapter 13 that the rate at which work is done is called *power* and is measured in watts (W). Recall also that power is calculated using the formula

$$P = \frac{W}{\Delta t},$$

where P is the power (in watts), W is the work performed (in joules), and Δt is the time (in seconds) that it took to do the work. Let's find your "personal" power.

E Go to the bottom of the staircase. Have a partner at the top of the staircase ready the stopwatch. He should say "1-2-3-Go!" and start the watch. When you hear "Go!" climb the staircase slowly.

F Your partner should stop the watch when you reach the top and stop climbing. Record the time in Table 2.

G Repeat steps E and F again, this time climbing the stairs at a normal pace. Record the time in Table 2.

H Do steps E and F again, this time climbing the stairs briskly. Do not run, jump stairs, or do anything unsafe! Record the time in Table 2.

> Calculate the power that your body delivered for each trip up the stairs. Use the work value that you calculated in Step D and the three different time values that you just obtained. Round your results to the nearest 1 W and record them in the appropriate rows of Table 2.

5. If you wanted to climb the staircase in half the time that it took you to climb it briskly, how much power would your muscles have to deliver?

_____

ENERGY VERSUS POWER

So which is more important, energy or power? There's no nice, neat answer. Scientists and engineers use both, depending on what they're doing or trying to describe. Let's look at a brief example.

Most of us think of explosives as "powerful," which is quite correct. One kilogram of TNT contains 4 700 000 J (4.7 MJ) of chemical energy. TNT detonates in just a fraction of a second, so it also produces a great deal of power since it releases lots of energy very quickly.

In contrast, one kilogram of butter contains 30 000 000 J (30 MJ) of chemical energy, more than six times as much as TNT! But the energy in butter cannot be released quickly. Butter has much more energy than TNT but can produce far less power.

6. Why is power more important than energy in an explosive?

_____
_____
_____
_____

7. Why is energy more important than power for a heating fuel?

_____
_____
_____
_____

WHAT ABOUT THE HORSES?

By now you may be wondering where horsepower (hp) fits into the picture. *Horsepower* was James Watt's original unit for measuring and expressing power. We're not sure how he worked out the value of a horsepower since there are several different stories.

According to one account, Watt measured the force (in lb) that a horse had to exert to turn a circular grinding mill. Next, he measured the distance that the horse walked in one minute (ft/min) as it turned the mill. Combining these results, he concluded that a typical horse could deliver 33 000 ft·lb/min. A steam engine doing the same work at the same speed would therefore deliver as much power as a horse. In other words, it would be a 1 hp engine.

The horsepower unit is no longer used in science. The SI unit of power, the watt, has replaced it. But as any sports car or monster truck enthusiast knows, the automotive industry still uses horsepower to rate engine power. In fact, electric motors and engines are still rated in horsepower in many countries.

Let's wrap up this activity by determining your "personal" horsepower. To convert from watts to horsepower, use the following conversion factor.

$$\frac{1 \text{ hp}}{746 \text{ W}}$$

8. How much horsepower do your muscles deliver when you walk up the staircase briskly?

_____

Table 1

Weight (N)	
Step Height (m)	
Steps	
Total Height (m)	
Work (J)	

Table 2

	TIME (s)	POWER (W)
Slow		
Normal		
Brisk		

Name
Date

CHAPTER 14: *ENERGY*

LAB 14B
TIME TO CLIMB

Designing a Better Motor

Going up a steep hill is hard work, even for a car. On a hill, not only must a car motor push the car forward—it must also lift the weight of the car a vertical distance equal to the height of the hill. It would be great if automotive engineers had access to car engines capable of unlimited power output, but that's unrealistic. Engineers must work within certain constraints, called *design parameters*. For cars this might mean being limited to a certain model of engine and to restrictions on vehicle size and weight. How can engineers improve a car's hill climbing potential while sticking to a prescribed set of design parameters?

To demonstrate how this design process works, you will be presented with a similar engineering problem: how to get the most work out of a particular engine design. The motor that you will use is called a *spool motor*, for reasons that you'll soon see. First you will build the motor, then you will be tasked with designing an improved version of it. To test the effectiveness of your design changes, your spool motor will be required to accomplish a certain task—climbing a hill.

QUESTIONS

How can I maximize the efficiency of an energy conversion?

- What kinds of energy conversion are shown by a spool motor?
- How can I improve a spool motor?

Time to Climb 145

PROCEDURE

Part 1: Building the Spool Motor

Refer to the diagrams below to build a spool motor.

A Use the wooden dowel to thread the rubber band through the spool. When you finish, there should be two loops left over, one at each end of the spool.

B Secure one end of the rubber band by inserting the wooden peg through one of the loops. Pull the rubber band from the opposite end to make the peg secure.

C Place the metal washer over the opposite end of the rubber band loop, then run about 1 cm of the wooden dowel or pencil through the loop. If the loop is loose, turn the spool until the connection is snug.

D Place a short piece of tape over the pegged end of the rubber band so that the peg cannot rotate.

Equipment
- wooden spool
- wooden dowel or pencil, approx. 20 cm
- wooden peg, approx. 1 cm
- metal washer
- rubber band
- tape

To operate your spool motor, twist the rubber band by turning the spool while holding the dowel. Do not release the spool and dowel until you are ready to run your motor! Use both hands to place your spool motor on a flat surface, then release it and watch it go!

1. What is the energy source of your spool motor?

   _____
   _____

2. Describe the energy conversion that takes place when the spool motor is released.

   _____
   _____

Part 2: Improving the Design

DESIGN PARAMETERS

Your task is to design an improved version of your spool motor that will be capable of transporting a load (the dowel) up a hill. You must work within the design parameters listed below.

- You may modify any part(s) of your spool motor with the following exceptions:
  » You must use the original rubber band from Part 1 of this lab activity or, in case of breakage, one exactly like it.

» You may not modify the length or mass of the dowel.

- You may add additional materials to your design so long as they do not violate the restrictions listed above.

- Your teacher will provide you with information about the steepness of the test hill that your improved motor must climb and tell you how far up the slope your motor must travel.

- Your teacher will show you the hill that your motor must climb.

3. How does not being allowed to modify the rubber band and dowel for your spool motor model what automotive engineers must do?

_____
_____
_____

4. Describe the energy conversion that must take place as your improved spool motor climbs the hill.

_____
_____
_____

5. On the basis of your answers to Questions 1–4, indicate exactly what it is that your ideally improved spool motor should be able to do.

_____
_____
_____

PLANNING THE DESIGN

E Keeping in mind the design parameters listed above, brainstorm possible improvements to your spool motor with your team.

F Each team member should individually research how the proposed improvements to the design or materials of the spool motor may affect its performance.

G Draw and label a diagram of your plan for an improved spool motor. Indicate on the diagram that this is a *preliminary design*. Label the spool motor's parts. Include a description of how any proposed changes to the original design are intended to improve the performance of the new design.

H Share your research findings and preliminary design with your team. Reach consensus on a final design that your group will build and submit for testing.

I Have someone in the group draw and label a diagram of the final design in the same manner as was done in Step G. Indicate on this diagram that it is an *initial production design*.

TESTING THE DESIGN

J Follow the procedure provided by your teacher for testing your improved spool motor.

REFINING THE DESIGN

K On the basis of your testing results, make modifications to your improved spool motor as needed. If changes are made, draw and label a new diagram indicating that this is the *final production design*.

RETESTING THE DESIGN

L On the due date for your improved spool motor, be prepared to submit your preliminary, initial production, and final production design, and to test whether your design can meet the required performance expectation. Record your motor's performance result(s) on your final design diagram (e.g., maximum distance traveled up the hill or height attained, in cm). Observe the design and operation of other spool motors designed by your classmates and be prepared to discuss the merits and drawbacks of the various designs that you observe.

CHAPTER 15: THERMODYNAMICS

LAB 15A

METAL MYSTERY

Inquiring into Specific Heat

You've spent all morning helping a neighbor with yard work, and now you are looking forward to a dip in the pool. You anticipate the cool refreshing water, but you also realize that you have to cross the sizzling-hot concrete to reach the water. Why is the concrete so hot, while the water is still cool? Both the concrete and water have been sitting in the hot sun all morning. It all has to do with specific heat.

In this lab activity, you will investigate the heat transfer between two water samples. Then you will design an experiment that will allow you to identify a metal on the basis of its specific heat.

QUESTIONS

How can thermodynamics help identify a metal?

- What does the value of specific heat tell me about a material?
- As something cools, where does the energy go?
- How can I identify a substance by how it transfers energy?

Equipment

laboratory balance
calorimeter
hot plate
graduated cylinder, 100 mL
thermometer
glass beaker, 250 mL
wire gauze
room temperature water
goggles
hot mitts (2)

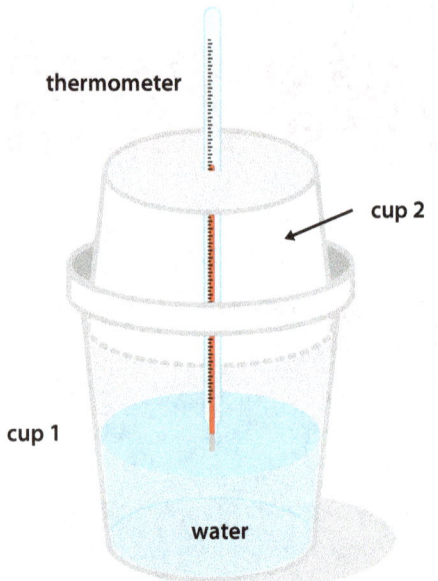

A calorimeter can be easily made with two foam cups and a thermometer.

PROCEDURE

1. What is specific heat?

_____
_____

The lower the specific heat of a material, the more its temperature will change as energy flows into or out of the material. On the other hand, a high specific heat results in less temperature change as energy is transferred. For example, water has a high specific heat (4.18 J/g°C). This means that water can absorb a large amount of energy with only a small change in temperature.

2. What would you assume about the specific heat of the concrete mentioned in the opener?

_____
_____
_____
_____

Understanding Energy Transfer

In this portion of the lab activity, you will use a calorimeter to understand the energy transfer between two samples of water. A calorimeter consists of a covered insulated cup with a hole for a thermometer. The purpose of the calorimeter is to study energy transfer within the system.

3. Why do you think that the calorimeter is insulated?

_____
_____
_____

A Mass the graduated cylinder and record this value in Table 1 (p. 152).

B Fill the graduated cylinder with 50 mL of room-temperature water. Mass the graduated cylinder and water and record your data in Table 1.

C Calculate the mass of the room temperature water and record this value in Table 1.

D Pour the water into the calorimeter.

E Measure the starting temperature of the room-temperature water and record your data in Table 1.

F Fill the beaker with 100 mL of water. Heat the beaker on the hot plate at medium-high heat. *Always have someone monitoring the heat source.* Continue heating until the water is approximately 60 °C. Turn off the hot plate.

150   Lab 15A

G Use the hot mitts and graduated cylinder to carefully measure 75 mL of the hot water. Place the beaker on the wire gauze.

H Measure the mass of the graduated cylinder and hot water and record this mass in Table 1.

I Measure the starting temperature of the hot water and record it in Table 1.

J Use the hot mitts to carefully pour the hot water into the calorimeter. Cover the calorimeter and insert the thermometer. Stir the water until the temperature stops changing. Record the final temperature of the mixture of the two volumes of water in Table 1.

K Calculate the mass of the hot water and record it in Table 1.

Analysis

4. What do you notice about the masses of the water samples when compared with the volumes of the samples? Explain the similarities and differences.

   _____
   _____
   _____
   _____
   _____

L Use the specific heat formula, $Q = c_{sp} m \Delta T$ (see page 359 in your textbook) to calculate the energy absorbed by the room-temperature water to warm it. Record your answer in Table 1.

5. Where did this thermal energy come from?

   _____

6. Write a hypothesis about the amount of energy lost by the hot water sample.

   _____
   _____
   _____
   _____

Metal Mystery

M Calculate the amount of energy that was lost by the hot water as it cooled. Record your answer in Table 1.

Table 1

	SPECIFIC HEAT (J/g°C)	MASS (g)	STARTING TEMPERATURE (°C)	FINAL TEMPERATURE (°C)	HEAT (J)
Graduated Cylinder					
Graduated Cylinder and Room-Temperature Water					
Graduated Cylinder and Hot Water					
Room Temperature Water	4.18				
Hot Water	4.18				

7. State a claim about your hypothesis. Support your claim with evidence from your data.

_____

_____

_____

Scientific Questions

How can we use the information from our previous investigation to develop an experiment that will allow us to identify a metal on the basis of its specific heat?

Designing Scientific Investigations

N Write procedures to collect data that will allow you to answer the question above.

O Have your teacher approve your procedures.

Conducting Scientific Investigations

P Following the procedures that you have written, collect the data to answer the scientific question.

Developing Models

Q Analyze your collected data to determine the specific heat of the metal that you are testing.

Table 2

METAL	SPECIFIC HEAT (J/g°C)
Aluminum	0.91
Brass	0.40
Copper	0.39
Iron	0.45
Steel	0.51
Tin	0.21
Zinc	0.39

R Use Table 2 to identify the metal that you are testing.

Scientific Argumentation

S On the basis of the results of your test, make a claim regarding specific heat and the metal that you tested. As always, support your claim with evidence from your experimental data.

Going Further

8. Consider the materials in Table 2. Explain why an engineer might choose to use aluminum versus tin to solve a particular problem.

9. Do you think that an engineer makes her material decisions solely on the basis of specific heat? Explain.

CHAPTER 15: THERMODYNAMICS

LAB 15B

AROUND THE CURVE

Investigating the Heating Curve of Water

We all know that as we heat a substance, its temperature rises. But does it always behave that way? The answer is no, and we all actually know that it doesn't. You may be starting to wonder, "Have I ever seen something being heated without its temperature also rising?"

Let's think about an ice-cold soda, sitting there on the table, filled with ice on a hot summer day. Obviously, since the room is warmer than the soda, energy must be moving from the room into the soda. But fifteen, or even thirty, minutes later, it is still ice-cold. There is less ice in the glass, but the soda is no warmer than it was at the start. How does that work?

In this lab activity, you will investigate the change in temperature of a beaker of ice water as it transitions from ice water to boiling water.

QUESTIONS

Where does the thermal energy go when ice melts and water boils?

- What happens when I heat a substance?
- Why does a heating curve look the way it does?
- Where does the energy go when a substance changes state?

Equipment

Labdisc temperature probe
Labdisc Gensci
laboratory balance
hot plate with magnetic stirrer
magnetic stirring bar
beaker, 400 mL
ring stand
ring stand clamp
ice
wire gauge
goggles
hot mitts (2)

PROCEDURE
Collecting the Data

1. What temperature would you expect ice water to be? Explain.

2. What happens to the temperature of the ice water (ice still present) as it is heated? Explain.

3. At what temperature would you expect water to boil? Explain.

4. Write a hypothesis regarding what will happen to the temperature of water as it boils.

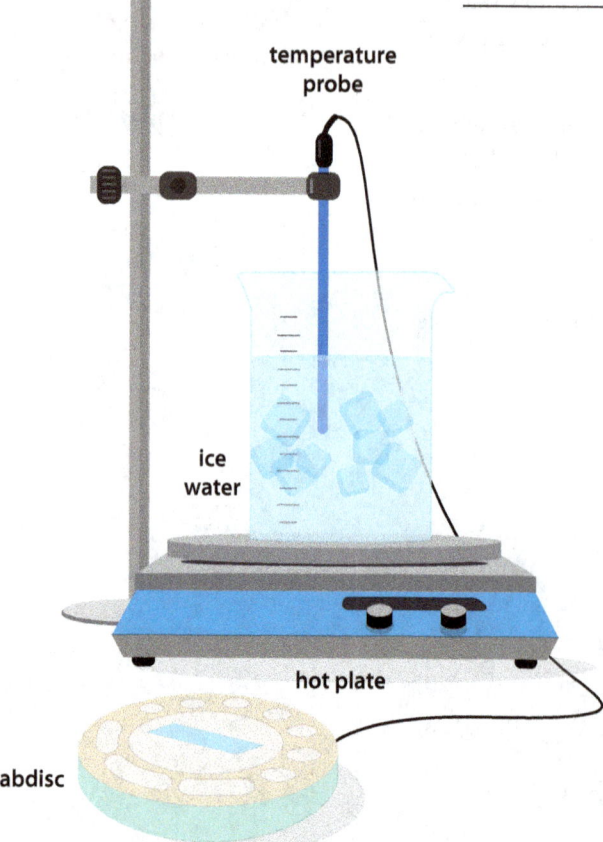

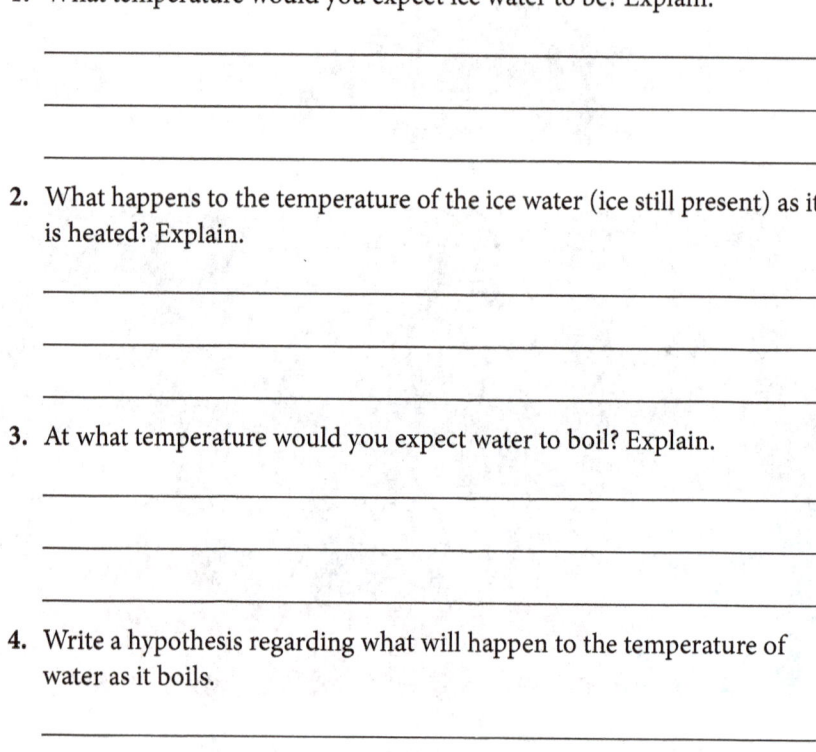

A. Plug in the temperature probe and turn on the Labdisc. (Refer to Appendix F for questions on use of the Labdisc.)

B. Set up the Labdisc to collect data from only the temperature probe.

C. Set the **Sampling rate** to **1/second**.

D. Set the **Number of samples** to **10 000**.

E. Make sure that the Labdisc is displaying *Extern temperature, °C*.

F. Place a magnetic stirring bar in the beaker.

G. Mass the beaker and stirrer. Record this mass in Table 1.

H. Fill the beaker halfway with ice, then add water to the 300 mL mark. Add water as needed to ensure that the stirrer bar will not strike the ice.

I. Mass the beaker, stirrer, water, and ice. Record this mass in Table 1.

J. Calculate the mass of just the water and ice. Record this mass in Table 1.

K. Place the beaker onto the hot plate (with the heat turned off).

L. Turn on the stirrer so that the bar just rotates.

M. Set up the ring stand and clamp so that it will hold the temperature probe in the center of the water in the beaker.

N. Monitor the temperature until the temperature stabilizes.

O. Turn the hot plate on high and start collecting data. *Always have someone monitoring the heat source.*

P. Note the temperature when all the ice has melted. Also note the temperature at which the water starts to boil.

Q. Once the water boils, continue heating and recording temperatures for two more minutes.

R. Stop data collection and turn off the hot plate.

S. Use the hot mitts to carefully move the beaker to the laboratory balance. Record the mass of the beaker, stirrer, and water in Table 1.

T. Use the hot mitts to carefully move the beaker to the wire gauge on the laboratory table.

U. The total mass changed because some of the water was converted into steam. Calculate this mass of the water converted to steam and record it in Table 1.

ANALYSIS

V. Start the Globilab software and download the last data set.

5. Describe the heating curve for water.

   _____
   _____
   _____

6. What happened to the temperature of the water while there was still ice in it? Where was the energy going if the temperature was not increasing?

   _____
   _____
   _____
   _____

W Use the specific heat formula to calculate the energy absorbed to heat water from 0 °C to 100 °C. Enter this value in Table 1. Show your work in the space below.

7. At what temperature did the water start to boil? Is this the temperature that you were expecting? If it is different, why do you think it is different?

___
___
___

8. What happened to the temperature as the water boiled? Where did the energy go if the temperature was not increasing?

___
___
___
___

X Calculate the energy to make steam by doing a unit conversion to change the mass calculated in Step U to joules (conversion factor: 2257 J/g). Record this value in Table 1. Show your work in the space below.

9. Why do you think it takes less energy to heat all the water than it takes to convert part of the water from water to steam?

___
___
___
___

10. Steam is used both for transporting thermal energy and for converting it to mechanical energy. On the basis of what you have learned, why do you think it is so good for these purposes?

_____

_____

_____

_____

_____

11. On the basis of what you have learned in Labs 15A and 15B, explain why it requires much more energy to turn liquid water into steam than to turn ice into liquid water.

_____

_____

_____

_____

_____

Table 1	
BEFORE HEATING	**MASS (g)**
Beaker and Stirrer	
Beaker, Stirrer, Water, and Ice	
Water and Ice	
AFTER HEATING	**MASS (g)**
Beaker, Stirrer, and Water	
Water Converted to Steam	
CHANGING STATE	**ENERGY (J)**
To Heat Water	
To Make Steam	

CHAPTER 16: FLUIDS

LAB 16A
HIGH PRESSURE JOB

Investigating Fluid Mass and Pressure

It's possible that you live not too far from a water tower. You may have noticed that they are very tall. And they are often placed on top of hills. Do engineers love tall towers, or is there a practical purpose to this design?

QUESTIONS

Why are water towers so tall?

- What is the relationship between water pressure and depth?

PROCEDURE

Water Height and Gravity

A Obtain a soda bottle having three small holes.

B Place the dowel rod alongside the holes and place pieces of strong tape over each hole and around the dowel.

1. Write a hypothesis about whether water pressure, when water is added to the bottle, will be the same or different at each of the three holes.

   _____
   _____

C Fill the bottle with water until the surface is more than 3 cm above the top hole.

D Set the bottle in a dishpan or sink. Uncover the holes simultaneously by pulling on the dowel.

Equipment

beakers, 600 mL (2)
laboratory tubing, 1 m (2 lengths)
soda bottles, 2 L (2)
wooden dowel, 30 cm
tape
dishpan
goggles

High Pressure Job   161

2. How does the pressure exerted by water differ at each hole, if at all?
_____
_____

3. Did your data support your hypothesis? Explain.
_____
_____

E Empty the soda bottle and dishpan into a sink.

F Fill the dishpan with approximately 900 mL of water and set it on the table.

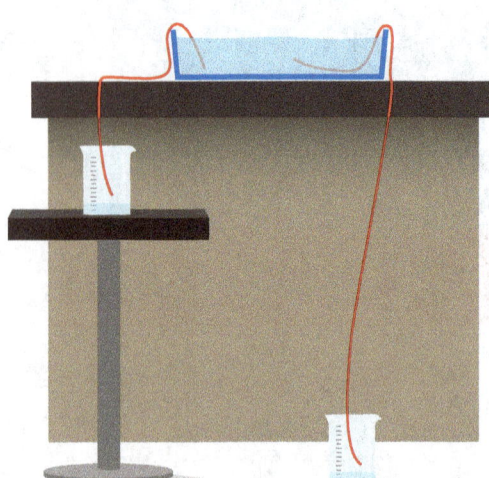

G Set one 600 mL beaker on the floor and the other on a chair.

4. Write a hypothesis about whether water flowing through laboratory tubing from the dishpan into these beakers will flow at the same or different rates?
_____
_____

H Fill two 1 m lengths of laboratory tubing by submerging them in water and place your thumbs over one end of each hose. Your lab partner should use his thumbs to close the other ends of the hoses.

I Have your lab partner place his ends of the tubes in each beaker. Place your ends in the dishpan and hold them near the bottom.

J You and your lab partner should remove your thumbs from the ends of the tubing but continue to hold the tubing in place. You have just created a siphon.

5. What occurred when you removed your thumbs from the ends of the tubing?
_____
_____

K When the water level in the dishpan becomes so low that air starts to run through the laboratory tubing, remove your ends of the tubing from the dishpan and hold them up until all the water has run out into the beakers.

6. How do the water levels in the two beakers compare?
_____
_____

7. Did your data support your hypothesis? Explain.
_____
_____

8. In each of these experiments, what force is causing the water to flow?
   _____

9. On the basis of your observations with the soda bottle, write a statement about the relationship between water pressure and the depth of a column of water.
   _____
   _____

10. The relationship between pressure and water column depth is also at work in the siphon. But the depth of the water column in the dishpan is the same for both beakers. Explain how the water column was deeper in the siphon to the beaker on the floor.
    _____
    _____

GOING FURTHER

Now let's go back to considering water towers. They are built with a large reservoir on top and a central pipe running down from the center of the reservoir. As you can see illustrated at right, the water column extends from the top of the reservoir to the lowest point in the water system.

11. Using what you learned in your experiments, why do you think water towers are so tall?
    _____
    _____
    _____

12. Theoretically, an electric pump could provide the necessary pressure to operate a water system. What advantage would the water tower have over a pump during a power outage?
    _____
    _____
    _____

13. Engineers have learned how to apply the principles of physics to improve water systems, but many of these engineers are unbelievers. Explain why a biblical worldview shows that a scientist's work is valuable even if his motives are not.
    _____
    _____
    _____
    _____
    _____

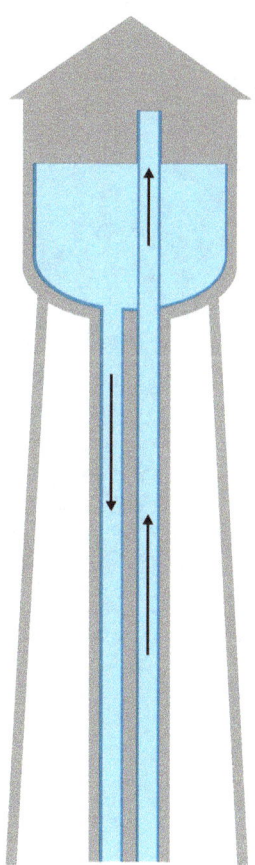

CHAPTER 16: FLUIDS

LAB 16B
LOAD, LOAD, LOAD YOUR BOAT

Designing a Paper Boat

Humans have used ships to move goods and people for millennia. Ships come in many different shapes and sizes, depending on their uses. To this day, everything from food to cars is transported on ships. Naval architects create different ship designs according to the ship's purpose. For instance, a cruise liner is quite different from a roll-on/roll-off (RORO) vessel because the cruise liner is designed to carry lots of people and amusements, while RORO vessels are designed to transport wheeled vehicles.

In this lab activity, you will be tasked with designing a boat that is optimized for carrying as much cargo as possible.

PROCEDURE

Design Parameters

As you design your boat, you must work within the design parameters listed below.

- You may use any boat design you want as long as it can be made from a single piece of 20 cm × 20 cm waxed paper.

- Your boat must be made from only the sheet of waxed paper and tape.

- Your goal is to maximize the number of pennies that your boat can hold. It does not need to be able to travel between two points.

QUESTIONS

How many pennies can a paper boat hold?

- What characteristics increase the buoyant force on a boat?

Equipment
wax paper sheets, 20 cm (2)
tape
pennies (250)
dishpan

Load, Load, Load Your Boat **165**

1. A paper boat holds 401 pennies. When the 402nd penny is added, the boat sinks. Why should the designer record that his boat held 401 pennies?

   _____
   _____

2. What determines whether an object floats or sinks in water?

   _____
   _____

3. Factors such as speed and maneuverability are not considerations for your design. How is this similar to what naval architects do when they design a vessel?

   _____
   _____
   _____
   _____

Planning the Design

A Keeping in mind the design parameters listed above, brainstorm possible designs for your paper boat with your team.

B Each team member should individually research how different possible designs may affect a paper boat's ability to hold pennies without sinking.

C Draw and label a diagram of your plan for your boat. Indicate on the diagram that this is a *preliminary design*.

D Share your research findings with your team. Reach consensus on a final design that your group will build and submit for testing.

E Have someone in the group draw and label a diagram of the final design in the same manner as was done in Step C. Indicate on this diagram that it is an *initial production design*.

Testing the Design

F Follow the procedure provided by your teacher for testing your boat.

4. A penny is denser than an equal volume of water. So why would a boat be able to hold multiple pennies and still float?

   _____
   _____

Name _____

Refining the Design

G On the basis of your testing results, make modifications to your boat, if needed. If changes are made, draw and label a new diagram indicating that this is the *final production design*. Include a description of how any proposed changes to the original design are intended to improve the performance of the new design.

Retesting the Design

H On the due date for your boat, be prepared to submit your preliminary, initial production, and final production designs and to test how your design measures up to the required performance expectation. On your final design diagram, record the number of pennies that your boat was able to hold. Observe the design and operation of other paper boats designed by your classmates and be prepared to discuss the merits and drawbacks of the various designs.

CHAPTER 17: PERIODIC MOTION AND WAVES

LAB 17A

TICK TOCK

Investigating Pendulums

Do you enjoy carnival rides? Have you ever ridden the pirate ship? Many children and adults enjoy this ride that swings back and forth. Many people do not realize it, but the pirate ship is a type of pendulum. The time that it takes for a pendulum to complete one full cycle is called its *period*. In this lab activity, you will experiment with the period of a pendulum.

QUESTIONS

What use is a pendulum?

- What is a pendulum?
- How do the properties of a pendulum affect its period?

PROCEDURE

A   Set up the ring stand and iron ring so that the pendulum will freely swing over the edge of the laboratory table.

B   Tie a loop at one end of the string, and hang the mass hanger on the loop. Hang a 50 g mass on the mass hanger.

C   Wrap the other end of the string several times around the iron ring. Pull the end of the string until the pendulum length—the distance between the bottom of the iron ring and the center of mass of the pendulum bob, the mass—is 70 cm.

Modeling Period Versus Mass

In this first part of the lab activity you will explore the relationship of a pendulum's period to its mass. You'll vary the mass while keeping the displacement (starting angle) and the pendulum length constant.

Equipment

ring stand
iron ring
mass hanger
mass set
meter stick
protractor
stopwatch
string (1 m)

Tick Tock   169

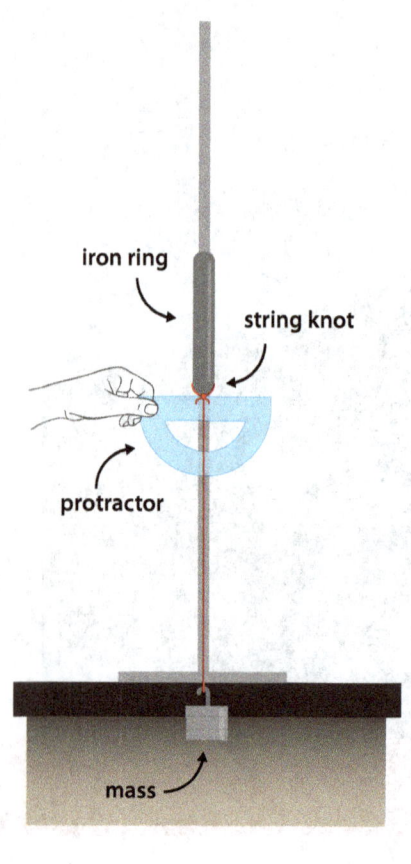

1. Write a hypothesis stating how you think the mass of the bob will affect the pendulum's period.

   _____

   _____

   D Place the straight edge of the protractor on the underside of the iron ring so that the string passes through the 90° mark. Pull the pendulum back 10°.

   E Have your partner count down to start the stopwatch. As he begins timing, release the pendulum and remove the protractor so that the string doesn't rub against it.

   The pendulum completes a cycle when it comes back to the place where you released it.

   F Count the pendulum cycles. Stop timing when the pendulum has finished ten complete cycles. Record the time for ten cycles in Table 1.

   G Divide the recorded time by 10 and record the period in Table 1.

2. Why would you time ten periods and divide by 10 instead of just timing one period?

   _____

   _____

   _____

   _____

3. What force made the pendulum swing back toward its equilibrium position?

   _____

   H Repeat Steps C–G for the other three masses listed in Table 1 by adding masses to the mass hanger. Be sure to adjust the string to maintain the pendulum length at a constant 70 cm.

   I Graph your data in Graphing Area 1. Add a curve of best fit.

4. Did the data support your hypothesis? Explain.

   _____

   _____

5. Why do you think the period behaved as it did in response to changes to the mass?

   _____

   _____

   _____

   _____

Name _____

Modeling Period Versus Displacement

In this second section of the activity, you will explore the relationship of the pendulum's period to its displacement (starting angle). You will vary the pendulum starting angle while keeping the pendulum length and bob mass constant.

6. Write a hypothesis stating how you think the pendulum's displacement will affect its period.

   _____
   _____

J Hang a 100 g mass on the mass hanger and adjust the string so that the pendulum length is 70 cm.

K Repeat Steps D–G for the four displacements listed in Table 1.

L Graph your data in Graphing Area 2. Add a curve of best fit.

7. Did the data support your hypothesis? Explain.

   _____
   _____

8. Why do you think the period behaved as it did in response to changes to the displacement?

   _____
   _____
   _____
   _____

Modeling Period Versus Length

In the third section of the activity, you will investigate the relationship of the pendulum's period to its length. You will vary the pendulum length while keeping the displacement and the pendulum mass constant.

9. Write a hypothesis stating how you think the pendulum's length will affect its period.

   _____
   _____

M Hang a 100 g mass on the mass hanger and adjust its string so that the pendulum length is 30 cm.

N Repeat Step D–G for the four string lengths listed in Table 1.

O Graph your data in Graphing Area 3. Add a curve of best fit.

Tick Tock   171

10. Did the data support your hypothesis? Explain.
   _____
   _____

11. Why do you think the period behaved as it did with changes to the length?
   _____
   _____
   _____
   _____

12. On the basis of what your data reveals, describe how changing the three properties you tested affects a pendulum's period.
   _____
   _____
   _____
   _____

GOING FURTHER

People commonly think of pendulums as part of grandfather clocks, but they are used for other purposes too. For example, very large pendulums are used to keep skyscrapers steady. Tall skyscrapers have a problem. On windy days, they sway back and forth. No one notices the problem on the ground floor, but at the top of the building the motion can be significant. In fact, some people in upper-floor offices and apartments have experienced motion sickness.

One way that engineers can solve this problem is to install a large modified pendulum called a *tuned mass damper*. Connected to the structure by giant springs, it swings as the building sways. This swinging absorbs much of the kinetic energy of the building, reducing the building's motion.

13. Where do you think the energy that the tuned mass damper absorbed goes?
   _____
   _____

14. Engineers could also reduce the amount of swaying a building does by using heavier materials or extra bracing. Why do you think they use a tuned mass damper instead?
   _____
   _____

Name _____

Table 1

	MASS (g)	DISPLACEMENT (°)	LENGTH (cm)	TIME FOR TEN CYCLES (s)	PERIOD (s)
VARYING MASS	50	10	70		
	100				
	150				
	200				
VARYING DISPLACEMENT	100	10	70		
		15			
		20			
		25			
VARYING LENGTH	100	10	30		
			50		
			70		
			90		

Graph Area 1

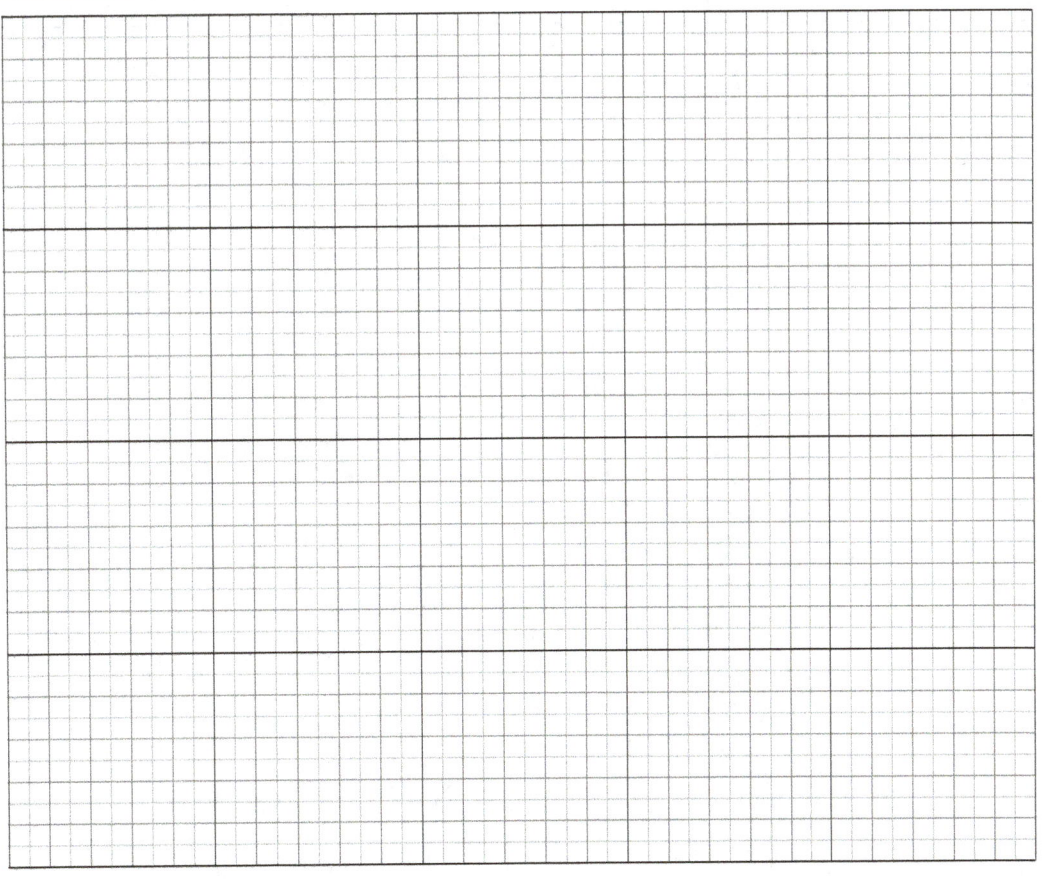

Tick Tock 173

Graph Area 2

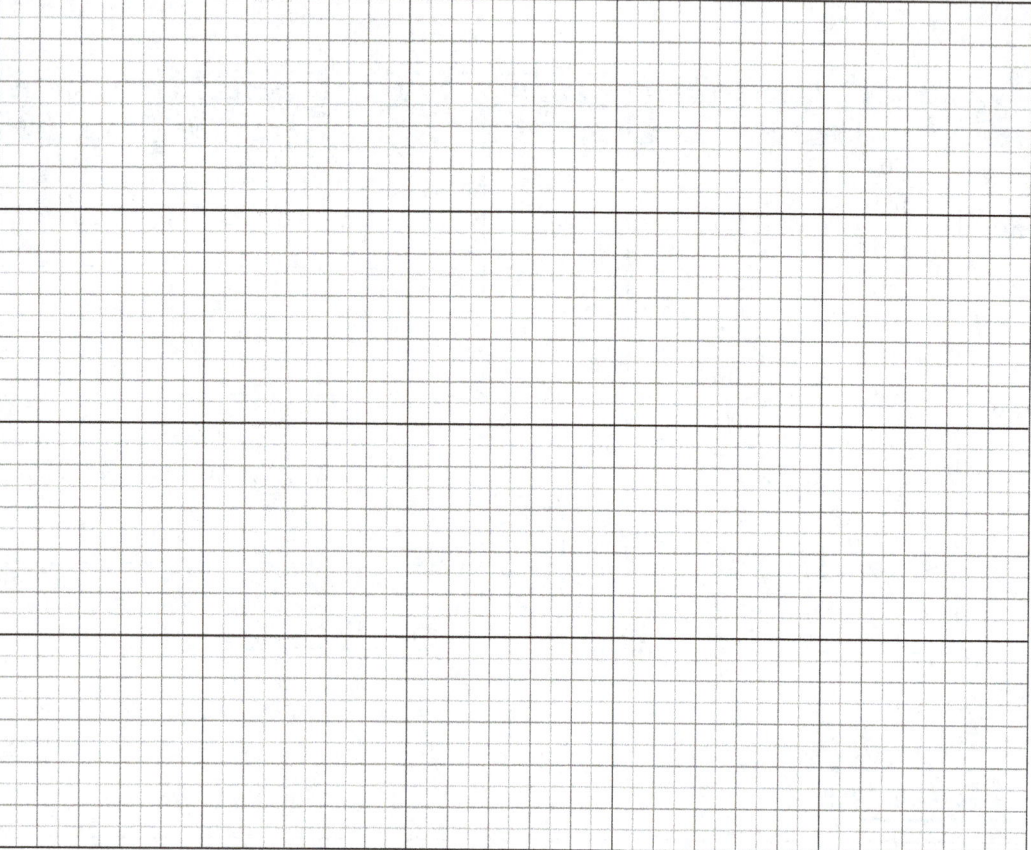

Graph Area 3

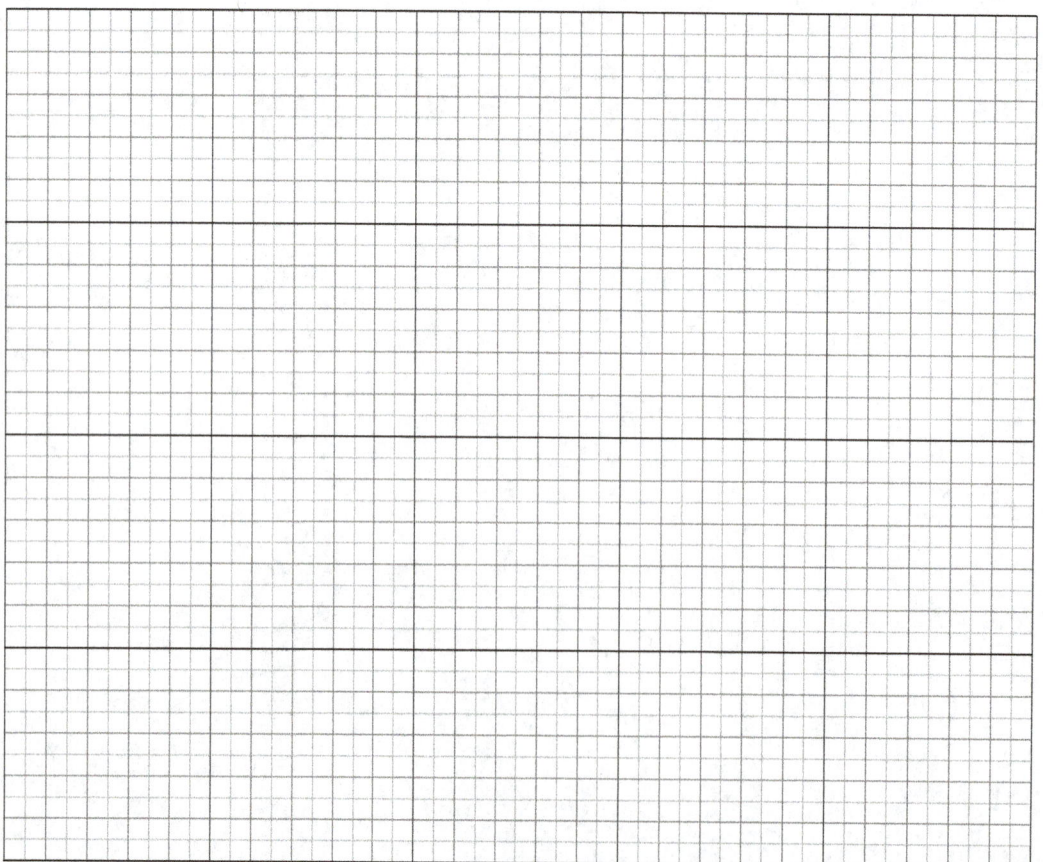

CHAPTER 17: PERIODIC MOTION AND WAVES

LAB 17B

STORM SURGE!

Creating Coastal Defenses

Many people love to live in coastal regions, but like all locations, there are drawbacks. A significant problem in many coastal regions is erosion. This problem is increased by storm surges. When storms blow in from the sea, the wind will often push the ocean water toward the shore, creating very large waves. In the worst hurricanes, these storm surges can penetrate several miles inland, causing severe flooding.

One way to help prevent flooding from storm surges is to build coastal defenses, but these present some difficulties. For one thing, they cannot be used on the entire coast because that would be too expensive. So coastal defenses are generally limited to population centers like cities or areas that experience the worst flooding. Another issue is that coastal defenses are often unsightly. This becomes an issue along recreational beaches visited by tourists. When engineers design coastal defenses, they must consider factors such as cost, aesthetics, effect on the environment, and impact on oceanic navigation.

QUESTIONS

How can I model coastal defenses?

- What are coastal defenses?
- How do coastal defenses counteract the effects of waves?
- Can effective coastal defenses be designed that allow ships to pass?

Equipment

graduated cylinder
large dishpan
plastic shelf bin
wave generator
toy boat

PROCEDURE

Design Parameters

Your task is to design a coastal defense that reduces the storm surge while not cutting off boat traffic to the shore. You must work within the design parameters listed below.

- Your materials are limited to the kinds and amounts indicated by your teacher.

- Your coastal defenses must allow two ships to pass at a time. Your teacher will provide you with information about the width of the vessels as well as necessary spacing between ships and between ships and the coastal defenses.

- Your teacher will show the unprotected shoreline by demonstrating the wave generator with no coastal defenses.

1. How does being limited in the kinds and amounts of materials model what engineers must do?

   _____
   _____
   _____
   _____

2. How does requiring that the coastal defenses allow two boats to pass at a time model what engineers must do?

   _____
   _____
   _____
   _____

Planning the Design

A Keeping in mind the design parameters listed above, brainstorm possible designs for coastal defenses.

B Each team member should individually research coastal defenses, balancing protecting the coast with allowing ship navigation.

C Draw and label a diagram of your plan for coastal defenses. Indicate on the diagram that this is a *preliminary design*.

D Share your research findings and preliminary design with your team. Reach consensus on a final design that your group will build and submit for testing.

E Have someone in the group draw and label a diagram of the final design in the same manner as was done in Step C. Indicate on this diagram that it is an *initial production design*.

Testing the Design

F   Follow the procedure provided by your teacher for testing your coastal defenses.

Refining the Design

G   On the basis of your testing results, make modifications to your coastal defenses. Draw and label a new diagram indicating that this is the *final production design*.

Retesting the Design

H   On the due date for your coastal defenses, be prepared to submit your preliminary, initial production, and final production designs. Be ready to test your design to see whether it meets the required performance expectations. Record your coastal defenses' performance result(s) on your final design diagram. Observe the design and operation of other coastal defenses designed by your classmates and be prepared to discuss the merits and drawbacks of the various designs that you observe.

CHAPTER 18: SOUND

LAB 18 A
SOUNDING OFF

Investigating the Properties of Sound

We live in a world filled with sounds. For this reason, we value our sense of hearing perhaps only slightly less than we do our vision. But how does sound really work?

Recall from your textbook that sound is a type of energy in the form of longitudinal waves moving through matter. When these waves enter our ears, we perceive them as sounds. But defining sound as waves falls a bit short of explaining the richness of our auditory experience. What causes sound to have so much variety?

The answer is that sound waves can be very complex. While science textbooks often use simple sound waves—waves for a single tone—to illustrate basic concepts, most real-world sounds are not simple waves. In this lab activity, we will look at the basic physical properties that give sounds their more obvious characteristics.

QUESTIONS

How are the different properties of sound related?

- Is there a relationship between wavelength, frequency, and pitch?
- Can I measure the speed of sound?
- Why do different materials conduct sound differently?
- How are sound waves reflected or bent?

Equipment
metric ruler (wood or plastic)
graduated cylinder, 100 mL
beaker, 50 mL
tuning fork
electric buzzer
megaphone
goggles

PROCEDURE

Wavelength, Frequency, and Pitch

One of the most obvious properties of a sound is its pitch—how high or low it seems to us. Our perception of pitch, though, is subjective; it's not a scientific dimension because it can't be measured. We call it a *perceptual phenomenon*, meaning that it depends on the individual listener's senses—*his* perception. Everyone hears the same sound slightly differently. That's one reason why some people are better musicians than others.

Since scientists try to avoid subjective data whenever possible, pitch is not very useful in a scientific setting. Instead, we must find a relationship between pitch and something that we can measure.

A Place the ruler so that it's projecting over the edge of a table with the 5 cm line along the edge. Firmly press the ruler down to the tabletop. Your fingers should be near the edge of the table (see left).

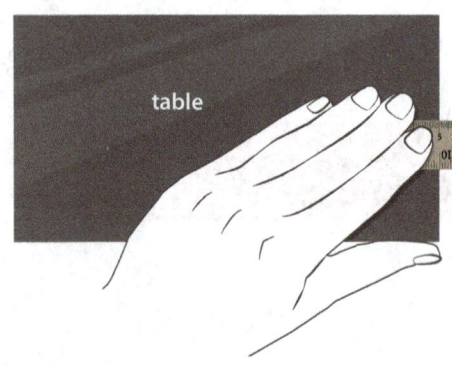

B Snap the free end of the ruler with your other hand. Listen to the pitch of the sound produced.

C Repeat Steps A and B four times, setting the ruler at the 10 cm, 15 cm, 20 cm, and 25 cm marks.

1. What happens to the sound's pitch as you shorten the vibrating end of the ruler?

   _____

2. What happens to the speed of the ruler's vibrations as the vibrating end is shortened?

   _____

As you shorten the ruler, the amplitude of the vibrations at the free end decreases and sound waves are produced closer together. Recall that the distance from the start of one complete wave cycle to the start of the next is called *wavelength*. It's measured in meters and is represented by the Greek letter *lambda* (λ) in equations.

3. If the waves move closer together as you shorten the ruler, what is happening to the distance from one wave cycle to the next (the wavelength)?

   _____

4. On the basis of your answers to Questions 1 and 3, what do you think is the relationship between wavelength and pitch?

   _____

   _____

The speed at which the ruler vibrates up and down is called its *frequency*. It's measured in cycles per second, or hertz (Hz), and is represented by the letter *f* in equations.

5. What happens to the sound's pitch as the ruler vibrates faster?
   _____

6. On the basis of your observations, what do you think is the relationship between frequency and pitch?
   _____
   _____

The speed of a sound wave in a uniform medium is a constant value. The relationship between the speed of sound, its wavelength, and its frequency is shown by the formula

$$v = \lambda f,$$

where v is the speed of sound in meters per second, λ is its wavelength in meters, and f is its frequency in hertz. Since the wave speed through air is basically constant, if one sound's frequency is higher than another's, its wavelength must be shorter to maintain the constant wave speed.

The Speed of Sound

Sound waves require a medium to carry them. So it's logical to assume that the medium influences the speed (v) at which the waves move. We usually assume that the medium is air. Under room-temperature conditions, the speed of sound in air is about 345 m/s.

You can measure the speed of sound waves in air using a very simple method. If you send a pure tone into a tube, a standing wave will result inside the tube when the length of the tube is one-fourth the tone's wavelength. When you find the proper tube length during the following procedure, the sound waves will reinforce each other and suddenly become much louder. This condition is called *resonance*. If you measure the length of the tube at resonance and multiply the result by 4, you will know the sound's wavelength.

You will use a tuning fork as the tone source and a graduated cylinder as a tube. By adding water to the cylinder, you can adjust its effective length until you find the resonance point.

D Follow your teacher's instructions and use the beaker to add the indicated amount of water to the graduated cylinder.

E Start the tuning fork vibrating by tapping it about halfway along its length on your knee or on a hard rubber surface.

F Hold the vibrating fork horizontally over the mouth of the graduated cylinder (see right). Note the loudness of the sound. **Be sure that the vibrating fork does not touch the graduated cylinder.**

G Using the beaker, add or remove small amounts of water and repeat Steps E and F until the sound is loudest. At that point, the cylinder is at resonance.

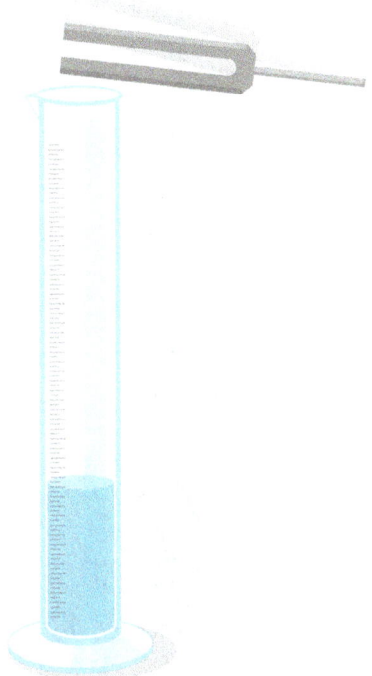

Table 1

Tube Length (m)	
Wavelength (m)	
Frequency (Hz)	
Speed (m/s)	

H Use the ruler to measure the distance from the mouth of the cylinder to the bottom of the water's meniscus—the tube length. Record your measurement in Table 1 on the left. Multiply this value by 4. This result is the wavelength of the tuning fork's sound waves. Record it in Table 1.

I The tuning fork's frequency should be stamped somewhere on the fork. Record this value in Table 1.

J Using the equation on page 181, calculate the speed of sound. Record this value in Table 1.

7. How close is your measured value for the speed of sound compared to the room-temperature standard value of 345 m/s?

   _____

8. Describe factors that may cause your measured value to be different from the room-temperature standard value.

   _____
   _____
   _____

Sound Transmission

A sound wave travels because a disturbance in some of the medium's particles creates a similar disturbance in the adjacent particles.

9. Hypothesize about what kind of medium transmits sound best, a solid or a gas.

   _____
   _____

K Have your partner tap a pencil's eraser on the table while you listen from a distance of 1 m. Note the loudness.

L Now have your partner tap the pencil's eraser on the table while you place your ear on the table 1 m from your partner. Note the loudness.

10. Was the sound louder when it passed through 1 m of air or 1 m of solid material?

    _____

11. Using the information provided at the beginning of this section, suggest an explanation for your answer to Question 10.

    _____
    _____
    _____

Reflection

Sound waves, like all waves, bounce off hard surfaces. This reflection is what causes echoes. A sound source emits sound waves in all directions if the wave path is unobstructed. If most of the sound waves can be made to travel in the same direction, the sound will be much louder.

M Have your partner sound the buzzer from a distance of 2 m. Note the loudness.

N Have your partner sound the buzzer in the small end of the megaphone while he points the large end *toward* you from a distance of 2 m. Note the loudness.

O Have your partner sound the buzzer in the small end of the megaphone while he points the large end *away* from you at a distance of 2 m. Note the loudness.

12. Using the information provided at the beginning of this section, suggest an explanation for what you've just observed.

_____
_____
_____
_____
_____

Diffraction

Do sound waves travel only in straight lines? If you think about it, you'll realize that sound waves must be able to bend around corners. After all, you're able to hear a conversation taking place around the edge of an open door. This behavior, known as *diffraction*, is something that all waves can do.

P Have your partner stand in the middle of a doorway and sound the buzzer. Stand at least 2 m away from your partner. Note the loudness.

Q Have your partner move to one side so that the wall is between him and you. You should stay in the same location. Listen as he sounds the buzzer again. Note the loudness.

13. Describe the buzzer's loudness when you heard it directly compared with when you heard it around the edge of the door.

_____
_____

14. Using the information provided at the beginning of this section, suggest an explanation for the difference in the sound levels.

_____
_____
_____

CHAPTER 18: SOUND

LAB 18B

SOUND ADVICE

Designing a Sound-Dampening Surface

Have you ever been to a NASCAR® race? While the speed of NASCAR fascinates most people, everyone remembers the noise—racecars roaring around the track, blasting their ears with sound! The crowd thunders as the winner crosses the finish line. Though this limited exposure to sound probably wouldn't damage hearing, think about the drivers and others who work at the track. Because of all the loud noise, NASCAR requires that workers wear hearing protection. Teams build their cars with insulation to protect drivers' hearing.

Hearing is one of our most important senses. You may not think much about protecting your hearing, but once it is gone, you can never get it back. That's why good employers think about hearing protection, especially since noise-induced hearing loss is the most common workplace injury in the United States. During this project, you will act like an engineer and design interior coverings to reduce the overall sound level in a simulated building.

QUESTIONS

What types of materials best reduce loud sounds in a building?

- How can surface materials affect sound levels?
- How can surface structure affect sound levels?
- How can surface texture affect sound reflection?

Equipment

test enclosure
tone-generating device
  (MP3 player, cellphone, tablet,
  or a speaker connected to one
  of these devices with a tone
  generator app)
sound measuring device (sound
  meter or cellphone with sound
  meter app)
surface coverings
meter stick

DESIGN PARAMETERS

Note: Record all notes, designs, data, calculations, analyses, and conclusions in a project log.

As you design your sound-reducing covering, you must consider both the structure and surfaces of the enclosure. Note the design parameters below:

- Your teacher will tell you the types of materials that you can use and the thickness limitations for your surface coverings. The walls and ceiling may have different parameters because more soundproofing material is often installed in ceilings than in walls.

- The surface coverings must fit within the dimensions of each wall area and the ceiling area of the enclosure.

- The surface coverings must fasten easily to the walls and ceiling of the enclosure.

- The surface coverings must consist of household materials or materials that you can purchase at a local home improvement store or discount department store.

PROCEDURE

Planning the Design

A Conduct research on materials for sound reduction as well as the effect of loud sounds on hearing.

B Measure the inside dimensions of the enclosure to determine the dimensions for your surface coverings.

C Draw and label your proposed design.

D Report your findings to your team members and discuss your options for wall and ceiling coverings. Reach a consensus on the materials and design that your team will use. Determine who will obtain materials and construct wall and ceiling coverings.

Testing the Design

Record all data in Table 1.

E Before you install the sound-reducing materials, measure the sound level as a benchmark so that you can determine the degree of sound reduction.

» Place the sound-measuring device in its place and begin recording.

» Place the tone-generating device in its place and turn it on.

» Install the roof of the enclosure.

» Allow the measuring device to record 15–30 s after putting the roof on.

» If you are recording with an app that records the data, download the data to a spreadsheet and produce a graph. If the app produces a graph, print it out.

» Repeat using three different tones (low, middle, and high frequency) for 15–30 s each.

F Install your surface coverings. Repeat Step E with the surface covering and record your data.

G Make any necessary modifications to your surface covering on the basis of your results from Step F, and then retest.

H Share your team's data with the other teams in your class.

I Compare your team's design with the other designs in a team discussion. Try to explain why the different designs performed the way that they did.

J On the basis of what you've seen, come to a conclusion about which materials and design provide the best soundproofing. Include any supporting information for your conclusion.

Table 1

DECIBEL LEVEL OF TONE	WITHOUT COVERING (db)			WITH COVERING (db)			DIFFERENCE (db)		
Team	low	medium	high	low	medium	high	low	medium	high

CHAPTER 19: ELECTRICITY

LAB 19A

GO WITH THE FLOW

Investigating Ohm's Law

Have you used anything made of electric circuits today? In our technology-driven world, it is hard to imagine going a day or even just a few minutes without using something that contains an electric circuit. Most of us use computers and phones throughout the day. Home lighting, heating systems, televisions, and entertainment systems all use electricity moving through circuits. Even just reading a book with a table lamp uses a circuit.

Circuits vary from the very simple, as in a lamp, to the super complex circuits of our smartphones and computers. All of these circuits are designed to control electrical energy so that it goes where it belongs and prevents it from going elsewhere. In this lab activity, you will build simple circuits to learn how current is controlled through a circuit.

QUESTIONS

How do resistance and voltage affect current in a circuit?

- What controls how much current moves through a circuit?
- How can I change the current in a circuit?

PROCEDURE

1. What is electrical resistance?
   _____
   _____

2. Compare open and closed circuits.
   _____
   _____
   _____

Equipment

multimeters (2)
power supply, variable DC
breadboard
jumper wires
resistors (5)

Go with the Flow   189

3. What do you think the colored strips on the resistors shown at left indicate?

_____

_____

Building a Circuit

A breadboard is a metal and plastic platform that serves as the foundation upon which a circuit can be built. It is designed to allow users to easily assemble and disassemble circuits. Notice the breadboard shown below. Understand that there may be variations in the way specific breadboards are marked, but the following description is typical of most breadboards.

The main portion of the breadboard is the configuration of terminal strips that form the two columns of rows of five holes down the center of the breadboard. Typically the terminal strips are in numbered rows, and the holes in each row are lettered. These holes are where connections are made between components of the circuit. Each row of five holes is connected by a metal strip beneath the plastic frame. For instance, the holes in Row 1, Columns a–e, are all connected to each other. The two columns of terminal strips are separated by a gap, which means that the holes in Row 1, Columns a–e, are not connected to the holes in Row 1, Columns f–j.

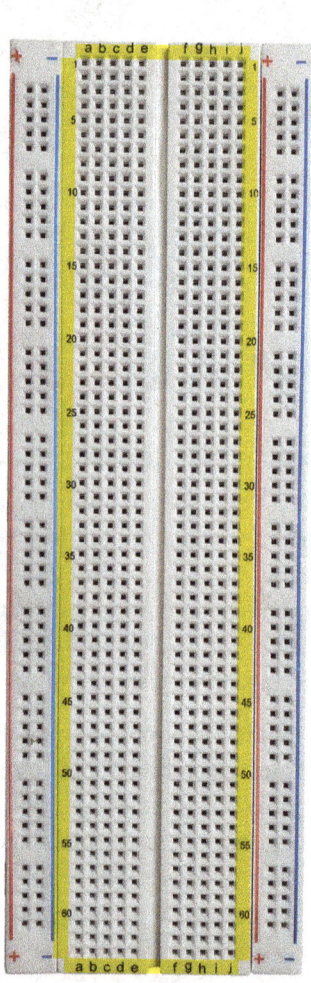

The four columns (two on either side) on the outer edges of the breadboard are the power rails. The two columns marked with a red line and "+" sign are positive terminals and should be connected to the positive terminal of the power supply. The columns marked with a blue line and a "–" sign are the negative or ground terminals and should be connected to the negative or ground of the power supply. The power rails are not connected to each other, though you could wire your circuit to connect them if needed.

A Set up one of the multimeters as an ammeter. Connect one of your leads to the *mA* port of the meter and the other lead to the common or ground port. Set the mode selector knob to 20 mA.

B Set up the other multimeter as a voltmeter. Connect one of your leads to the *V* port of the meter and the other lead to the common or ground port. Set the mode selector knob to 20 V DC.

C Set the power supply to 4.5 V DC.

D Use a jumper wire to connect the positive terminal of the power supply to the top hole in the "+" power rail on the left side of the breadboard.

E Connect the *COM* lead from the ammeter to the negative terminal of the power supply. Leave the other lead of the ammeter free until Step J.

F Insert one end of a jumper wire into 20b. Leave the other end of this jumper wire free over the edge of the breadboard.

A resistor is designed to convert electric energy into thermal energy. In an electric circuit, the purpose of a resistor is to control either current or voltage, or both. In some applications, the sole purpose of the resistor is to convert energy. The filament in an incandescent light bulb is a resistor that converts electric energy into thermal energy to the point that the filament begins to emit light.

G Use the resistor code chart at right to decode the resistance of your first resistor and record in Table 1.

H Insert one wire of the resistor into one of the holes in the "+" power rail on the left side of the breadboard. Insert the other wire into 20a.

Testing Current Versus Resistance

I Turn on power to the power supply.

J Touch the *mA* lead of the ammeter to the free end of the jumper wire connected to 20b. ***If any part of the circuit smokes or appears to be overheating, remove the ammeter lead from the jumper wire, turn off the power supply, and notify your teacher.***

K Read the value on the ammeter and record the current reading in Table 1.

L With the circuit still connected, touch the two leads from the voltmeter to the two exposed wires on either side of the resistor. Record the voltage reading in Table 1.

M Remove the ammeter lead from the jumper wire to open the circuit and turn off power to the power supply.

N Remove the resistor from the circuit.

4. Write a hypothesis about what you expect will happen to the current if the resistance is changed.

_____

_____

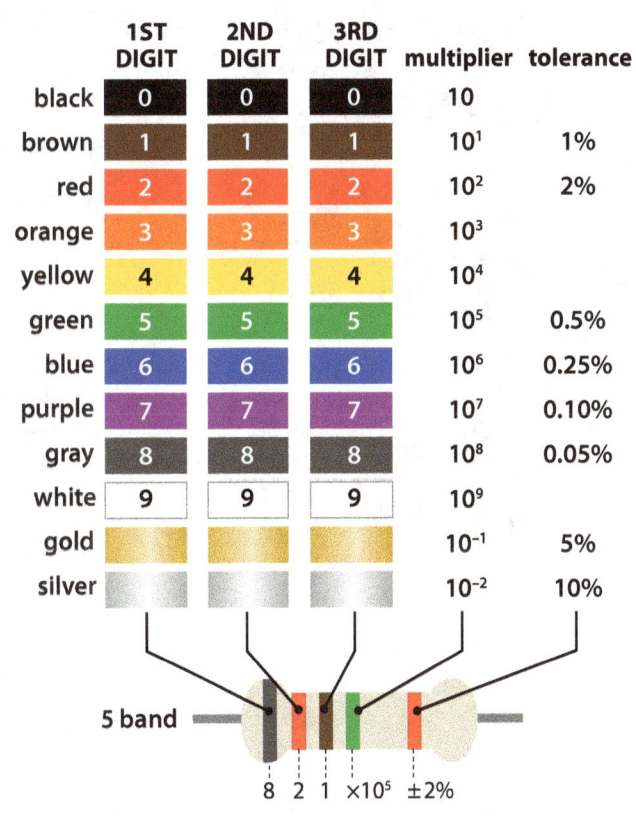

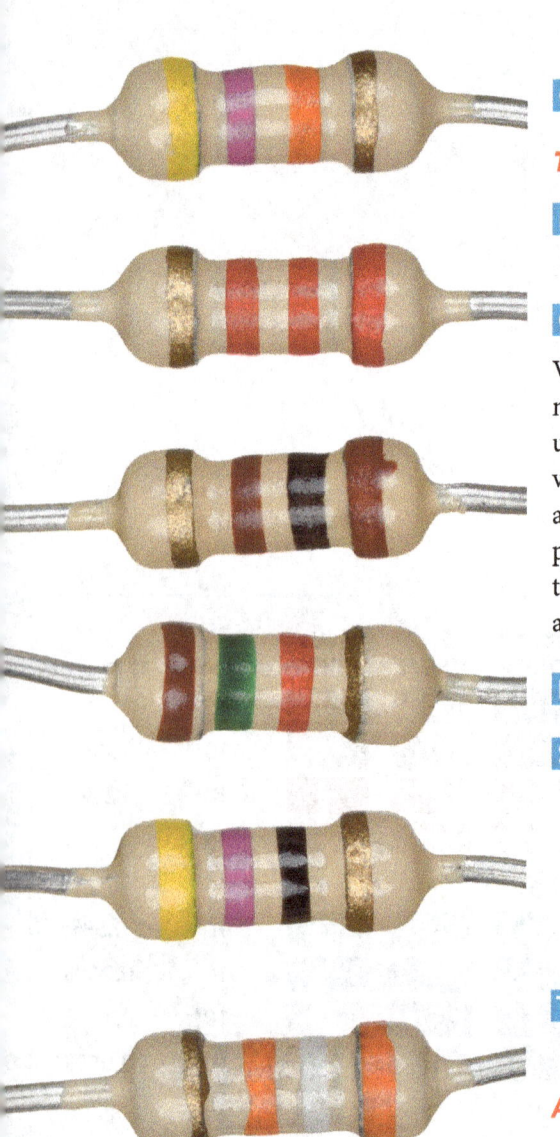

O Repeat Steps G–N for the remaining four resistors.

Testing Current Versus Voltage

P Replace the resistor in your circuit with the middle-value resistor and record this value in Table 1.

Q Turn on power to the power supply.

Voltage is the "push" that forces charge carriers through the circuit. It is the measure of the power source of the circuit. Most of the battery cells that we use (AA, AAA, and C cells) are sources of DC power and have an output voltage of 1.5 V. Our homes are powered by AC electrical systems and operate at 120 V. The power supply applies voltage to the circuit, and each component within the circuit uses some of that voltage "push" to do its work. The total voltage used by all the components must be equal to the original voltage applied to the circuit.

R Adjust the power supply voltage to 1.5 V.

S Repeat Steps J–K to measure the current and record in Table 1.

5. Write a hypothesis about what you expect will happen to the current if the voltage is changed.

_____

_____

T Repeat Steps R (adjusting the voltage each time per Table 1) and S for the remaining four voltages listed in Table 1.

Analysis

6. State a claim about whether your data supports your hypothesis about current and resistance (Question 4). Support your claim with evidence from your data.

_____

_____

_____

_____

7. State a claim about whether your data supports your hypothesis about current and voltage (Question 5). Support your claim with evidence from your data.

_____

_____

_____

_____

Going Further

8. How does the equation for Ohm's law below describe the relationship between current and both voltage and resistance? Does your data agree with the formula?

$$I = \frac{V}{R}$$

_____

_____

_____

9. You read earlier that the filament in an incandescent light bulb is a resistor. Can you think of another example of a resistor being used solely for the purpose of converting electric energy into thermal energy?

_____

_____

_____

Table 1

EFFECT OF CHANGING RESISTANCE		
RESISTANCE (Ω)	CURRENT (mA)	VOLTAGE (V)

EFFECT OF CHANGING VOLTAGE		
VOLTAGE (V)	RESISTANCE (Ω)	CURRENT (mA)
1.5		
3.0		
4.5		
6.0		
7.5		

CHAPTER 19: ELECTRICITY

LAB 19 B

SERIES-OUSLY?

Investigating Series Circuits

The term *series*—in connection with talking about a book series, television series, or movie series—is familiar to us all. We read or watch these in order so that the events in each book or show make sense. The word is also used in the context of electric circuits.

In Lab 19A you used a simple circuit to investigate Ohm's law. Through that lab activity you should have discovered the relationships between current and resistance and between current and voltage. As the resistance in a circuit increases, the current decreases, and as the resistance decreases, the current increases. Conversely, as the voltage increases, the current also increases, and as the voltage decreases, the current also decreases.

Most circuits that we use are more complex than the ones used in Lab 19A and involve many components that are connected. These components can be connected either *in series* or *in parallel*. Like the books in a series, the electricity moving through components connected in series moves through each component in sequential order. This means that there is only one path for the electricity to flow through in series components.

In this lab activity, you will expand on what was learned in Lab 19A. You will investigate the effect on current and voltage when additional components are added in series.

QUESTIONS

How does connecting components in series affect circuits?

- What is a series circuit?
- What happens to the voltage in components that are wired in series?
- What happens to the current in components that are wired in series?

1. Explain how the water flowing past the two bridges shown at left is like a series circuit.
   _____
   _____
   _____

2. Why do you think an engineer would wire two things in series?
   _____
   _____
   _____

Equipment

multimeters (2)
power supply, 6.0 V
breadboard
jumper wires
resistors, 220 Ω (3)
LEDs (3)

PROCEDURE

A Set up one of the multimeters as an ammeter. Connect one of your leads to the *mA* port of the meter and the other lead to the common or ground port. Set the mode selector knob to measure amperage. If your multimeter has different amperage settings, select 20 mA.

B Set up the other multimeter as a voltmeter. Connect one of your leads to the V port of the meter and the other lead to the common or ground port. Set the mode selector knob to DC V. If your multimeter has different voltage ranges, select 20 V.

C Turn on the power supply and set it to 6.0 V DC. Then turn the power supply off while you set up your circuit.

Testing Series Components

D Connect a lead wire to the positive terminal of the power supply and then use a jumper wire to connect the lead wire to the top hole on the "+" power rail on the left side of the breadboard.

E Connect the *COM* lead from the ammeter to the negative terminal of the power supply. Leave the other lead of the ammeter free until Step J.

F Insert one wire of one of the resistors into a hole on the "+" power rail on the left side of the breadboard. Insert the other wire into 10d.

G Insert the long wire from one of the LEDs into 10e and the short wire into 17e.

The resistor and LED are each a separate component. For the purpose of this lab activity, we will treat them as one.

The long and short wires of the LEDs are important because an LED is designed for current in one direction only. You may damage the LED if you wire it backward.

H Insert one end of a jumper wire into 17a and leave the other end extending over the side of the breadboard.

I   Turn on power to the power supply.

J   Touch the *mA* lead of the ammeter to the free end of the jumper wire connected to 17a. **If any part of the circuit smokes or appears to be overheating, remove the ammeter lead from the jumper wire, turn off the power supply, and notify your teacher.**

K   Read the value on the ammeter and record this reading in the first row of the *Current in Resistor* and *Total Current* columns of Table 1.

L   With the circuit still connected, touch the one lead from the voltmeter to the exposed wire on the side of the resistor near the power rail and the other lead to the exposed wire on the side of the LED near 17e. Record the voltage reading in the first row of the *Voltage in Resistor* and *Total Voltage* columns of Table 1.

M   Does the LED appear bright? Record the brightness of the LED in Table 1.

N   Remove the ammeter's *mA* lead from the jumper wire to open the circuit and turn off power to the power supply.

3. What do you think will happen to the current and voltage values if another resistor and LED are added to the circuit? What do you think will happen to the brightness of the LEDs?

_____

_____

O   Insert the free end of the jumper wire attached to 17a to 27a.

P   Add a series component by adding a resistor between 27b and 36b, another LED between 36c (long wire) and 40c (short wire), and another jumper wire at 40a with its other end extending over the side of the breadboard.

Q   Turn on power to the power supply.

R   Touch the *mA* lead of the ammeter to the free end of the jumper wire connected to 40a. Record this reading in the *Current in Resistor* column (second and third rows) and in the *Total Current* column of Table 1.

Remember that voltage is the "push" that causes the current in the circuit. When you measure voltage in different parts of the circuit, you should think of the voltages in slightly different ways. If you measure the voltage of a power supply, you can think of this as the "push" being supplied to the circuit. On the other hand, if you measure the voltage of a component in the circuit, you should think of this as the voltage used in that component. All the voltage put into a circuit by the power supply must be used by the components of the circuit.

Series-ously?   **197**

S With the circuit still connected, touch one lead from the voltmeter to the exposed wire of the first resistor on the side near the power rail and the other lead to the exposed wire on the side of the first LED near 17e. Make sure that the two devices are in between the voltmeter leads. Record the voltage reading in the second row of the *Voltage in Resistor* column of Table 1.

4. How did the voltage used in the first resistor-LED pair compare with the voltage used in the resistor-LED pair in the single-resistor circuit?

   _____

   _____

T Repeat Step S for the second resistor and LED.

U Repeat Step S but have the leads on either side of all four components (2 resistors and 2 LEDs). Enter this value in the *Total Voltage* column of the *2 Resistors* row of Table 1.

5. If 6.0 V was put into the circuit but only half the voltage was used in the second resistor-LED pair, where did the rest of the voltage go?

   _____

   _____

V Notice the brightness of the LEDs. Record the brightness of the LEDs for these in Table 1.

W Remove the ammeter lead from the jumper wire to open the circuit and turn off power to the power supply.

X Insert the free end of the jumper wire attached to 40a to 50a.

Y Add a third component in series by adding another resistor between 50e and 56e, an LED between 56c (long wire) and the 59c (short wire), and another jumper wire at 59a with its free end extending over the side of the breadboard.

Z Repeat Steps Q–W for the three resistors-LED pairs.

Analysis

6. How did the current through the first and second resistor-LED pairs compare in the 2-resistor circuit? Explain.

   _____

   _____

   _____

7. What happened to the total current in the circuit as resistor-LED pairs were added in series with each other? Explain.

___

8. What happened to the brightness of the LEDs as more resistor-LED pairs were added in series? Explain.

___

9. How did the total voltage used across all resistor-LED pairs compare with the voltage supplied by the power supply? Explain.

___

10. How did the voltage used across each resistor-LED pair compare with the voltage supplied by the power supply? Explain.

___

Going Further

If you have ever helped with Christmas lights, you may have been tasked with trying to figure out why an entire line of lights wasn't working. You worked tirelessly until you found the one bulb that was burned out. As you replaced the bulb, the entire line blinked to life.

11. Using what you have learned in this lab activity, how can you tell that the Christmas lights must have been wired in series?

___

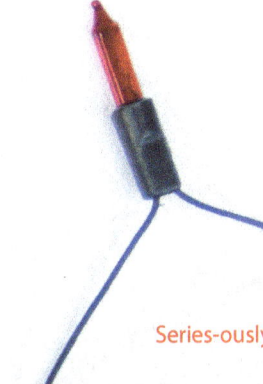

12. Instructions packaged with Christmas lights often warn users to limit the number of strings connected to three. If the lights are wired in series, why do you think this warning is given?

_____

_____

_____

_____

Table 1 EFFECT OF SERIES CIRCUIT

NUMBER OF RESISTORS	RESISTOR	INPUT VOLTAGE (V)				
		\multicolumn{5}{c}{6.00}				
		CURRENT IN RESISTOR (mA)	TOTAL CURRENT (A)	VOLTAGE IN RESISTOR (V)	TOTAL VOLTAGE (V)	BRIGHTNESS
1	1					
2	1					
	2					
3	1					
	2					
	3					

CHAPTER 19: ELECTRICITY

LAB 19C
THE PATH LESS TRAVELED

Investigating Parallel Circuits

Can you imagine if the plumbing in your house were connected in series? If you wanted to take a shower, you would have to go downstairs to the bathroom and turn on the faucet in the sink. Then you would go to the kitchen and turn on the water there and work your way up to your bathroom, turning on all the in-between faucets. Then you would finally get the water to pass through all those pipes and reach your shower.

Happily, that isn't how the plumbing is connected in our homes. We can use any plumbing fixture in the house whether or not any other fixture has water flowing. The plumbing fixtures in a house are connected *in parallel*. Components in an electric circuit can be connected in parallel too. While Lab 19B investigated the effect of connecting components along a single pathway, that is, in series, in this lab activity you will investigate the effect on current and voltage when additional components are added in parallel. The current in a parallel circuit has a choice of paths to take through components.

QUESTIONS

How does connecting components in parallel affect circuits?

- What is a parallel circuit?
- What happens to the voltage in components that are wired in parallel?
- What happens to the current in components that are wired in parallel?

The Path Less Traveled   201

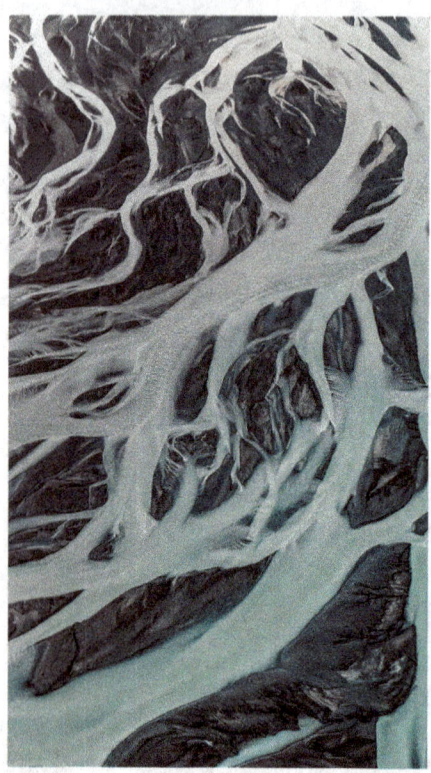

1. Use the river in the image at left as an analogy for parallel circuits.
   _____
   _____
   _____

2. Why might someone connect two things in parallel?
   _____
   _____
   _____

3. Think about the definitions of parallel and series circuits. Consider how lights and appliances work in a home. If a lamp has two bulbs in it and one burns out, the other light stays lit. You can also run a microwave in your home without needing to run the coffee maker. Are our homes wired in parallel or series? Explain.
   _____
   _____
   _____
   _____

Equipment
multimeters (3)
power supply, 6.0V
breadboard
jumper wires
resistors, 220 Ω (3)
LEDs (3)

PROCEDURE

A Set up two of the multimeters as ammeters. Connect leads to the *mA* port of each meter and other leads to the common or ground port. Set the mode selector knobs to measure amperage. If your multimeters have different amperage settings, select 200 mA on Ammeter 1 and 20 mA on Ammeter 2.

B Set up the third multimeter as a voltmeter. Connect one of your leads to the *V* port of the meter and the other lead to the common or ground port. Set the mode selector knob to DC V. If your multimeter has different voltage ranges, select 20 V.

C Set the power supply to 6.0 V DC, then turn it off while you set up your circuit.

Testing Parallel Components

D Connect a lead wire to the positive terminal of the power supply and then use a jumper wire to connect the lead wire to the top hole on the "+" power rail on the left side of the breadboard.

E Connect the *COM* lead from Ammeter 1 to the negative terminal of the power supply. Leave the other lead of the ammeter free until Step K.

F Insert one wire of one of the resistors into a hole on the "+" power rail on the left side of the breadboard. Insert the other wire into 10d.

G Insert the long wire from one of the LEDs into 10e and the short wire into 17e. *The long and short wires of the LEDs are important because an LED is designed for current in one direction only. You may damage the LED if you wire it backward.*

H Insert one end of a jumper wire into 17a and the other end into any hole on the "–" power rail on the left side of the breadboard. *Make sure that you insert it into the negative side of the power rail.*

I Connect two jumper wires to the "–" power rail. Connect one in the fifth-to-last hole and the other in the last hole. Let their free ends hang over the edge of the breadboard.

J Turn on power to the power supply.

K Touch the *mA* lead of Ammeter 1 to the free end of the jumper wire connected to the last hole of the "–" power rail. *If any part of the circuit smokes or appears to be overheating, remove the ammeter lead from the jumper wire, turn off the power supply, and notify your teacher.*

L Read the value on Ammeter 1 and record this reading in the first row of the *Current in Resistor* and *Total Current* columns of Tables 1.

M With the circuit still connected, touch one lead from the voltmeter to the exposed wire on the side of the resistor near the power rail and the other lead to the exposed wire on the side of the LED near 17e. Record the voltage reading in the first row of the *Voltage in Resistor* and *Total Voltage* columns of Table 1.

N Notice the brightness of the LED and record your observation in Table 1.

O Remove Ammeter 1's lead from the jumper wire to open the circuit and turn off power to the power supply.

4. What do you think will happen to the current and voltage values if a resistor is added to the circuit in parallel, giving the current two paths through which it may go? What do you think will happen to the brightness of the LEDs?

_____

_____

_____

P Add a second resistor by inserting one end in any hole on the "+" power rail and the other end in 23b. Add another LED between 23c (long wire) and 28c (short wire). Add a jumper wire at 28a with its other end connected to the "–" power rail.

Q Turn on the power supply.

R Connect the *mA* lead from Ammeter 1 to the jumper wire in the last hole of the "–" power rail. Both LEDs should be illuminated. Record the ammeter reading in the *Total Current* column for the *2 Resistors* row of Table 1. Also record the brightness of the lamps.

5. What happened to the total current in the circuit when you added the second resistor-LED pair in parallel?

    _____
    _____

S Remove the end of the 17a jumper wire from the "–" power rail. The first LED should go out. Connect the leads from Ammeter 2 to this jumper wire and the jumper wire connected to the fifth-to-last hole on the "–" power rail. Record this amperage in the *Current in Resistor* column for *Resistor 1* in the *2 Resistors* row of Table 1. Remove Ammeter 2 and reconnect the 17a jumper wire to the "–" power rail. Both LEDs should be back on.

T Repeat Step S for the second resistor-LED pair by using the jumper wire at 28a.

U Use the voltmeter as you did with the series circuits in Step M to measure the voltage in each of the resistor-LED pairs and also the voltage through both pairs together. Record the voltage readings in the appropriate columns of the *2 Resistors* row of Table 1.

6. How did the voltage used in each resistor-LED pair compare with the voltage put into the circuit? Explain.

    _____
    _____
    _____
    _____

V Turn off the power supply.

W Add a third resistor between the "+" power rail and 46c, a third LED from 46e (long wire) to 50e (short wire), and another jumper wire between 50a and the "–" power rail.

X Turn on the power supply. Connect the *mA* lead from Ammeter 1 to the jumper wire in the last hole of the "–" power rail. All three LEDs should be lit. Record the amperage from Ammeter 1 in the *Total Current* column in the *3 Resistors* row of Table 1. Also record the brightness of the lamps.

Y Repeat Step S for each of the three resistors in the circuit.

Z Use the voltmeter to measure the voltage for each of the resistor-LED pairs and also the voltage through all three pairs together. Record the voltage readings in the appropriate columns of the *3 Resistor* row of Table 1.

AA Turn off the power supply and disassemble the circuit.

Analysis

7. How did the current in each of the resistor-LED pairs compare with each other? Explain.

   _____
   _____
   _____
   _____

8. What happened to the total current in the circuit as resistor-LED pairs were added in parallel with each other? Explain.

   _____
   _____
   _____
   _____

9. How did the voltage used across each resistor-LED pair compare with the voltage supplied by the power supply? Explain.

   _____
   _____
   _____
   _____

10. What happened to the brightness of the LEDs as more resistor-LED pairs were added in parallel? Explain.

    _____
    _____
    _____

11. In Lab 19B, resistors were added in series, increasing the total resistance in the circuit, which resulted in decreasing the total current in the circuit. In this lab activity, as you added components in parallel, the total current increased. What can you conclude about the total resistance of the circuit as parallel components are added?

_____

_____

Going Further

Houses are wired in parallel, allowing people to have multiple appliances on a circuit with each one operating independently.

12. Using your data from this lab activity, suggest a problem that can occur when operating multiple appliances that are wired in parallel.

_____

_____

_____

Builders add safety features to the electrical systems in houses. Fuses or circuit breakers are added to each circuit in the house. Ground-fault interrupter outlets are also utilized in locations near water, like kitchens, bathrooms, garages, and outdoor outlets. Each of these items adds to the cost of building the house.

13. Justify the inclusion of safety equipment in homes and their associated costs.

_____

_____

_____

Table 1 EFFECT OF PARALLEL CIRCUIT

INPUT VOLTAGE (V)		6.00				
NUMBER OF RESISTORS	RESISTOR	CURRENT IN RESISTOR (mA)	TOTAL CURRENT (A)	VOLTAGE IN RESISTOR (V)	TOTAL VOLTAGE (V)	BRIGHTNESS
1	1					
2	1					
	2					
3	1					
	2					
	3					

CHAPTER 20: *MAGNETISM*

LAB 20A

LINES OF FORCE

Exploring Magnetic Fields

Magnets can be fun! You've probably seen them attract and repel each other even without touching. Scientists have gotten modern magnets to do some amazing things, including levitation. That's right—objects can levitate above a sufficiently strong magnet. In fact, scientists have levitated water, frogs, and mice.

At first, a magnet's ability to exert a force at a distance puzzled scientists. Then English scientist Michael Faraday proposed that magnetism radiates into space as curved lines of force. He called this region of influence a *field*. In this lab activity, you will investigate the lines of force in a magnetic field.

QUESTIONS

How can I find the shape of a magnetic field?

- What are some characteristics of the magnetic field around a bar magnet?
- Are the shapes of the magnetic fields between like and opposite magnetic poles different?

PROCEDURE

Measuring Field Strength Regions

A Cut a small slip of notebook paper the same length as one of your bar magnets and then tape it to one side of the magnet.

B Using a pencil, draw lines on the paper to divide the bar magnet into five equal sections. Starting at one end, label the sections A, B, C, D, and E.

C Lay out thirty paper clips in a loose collection and set the magnet on top. Lift the magnet and observe the location of the paper clips.

1. Describe how the paper clips were distributed over the magnet.

_____

Equipment
bar magnets (2)
small paper clips (1 box)
small magnetic compasses (5)
cardboard 20 cm × 20 cm
furring strips, 22 cm (2)
cardstock sheet 8.5 × 11
iron filings

2. On the basis of your observations, where on the bar magnet is the field strongest? Where is it weakest?

_____

_____

Determining Field Direction

Arrows drawn parallel to the field lines are used to indicate their direction. Scientists have agreed that the direction of a magnetic field (the way the field lines point) is the same direction that the north end of a magnet suspended in the field would point. Their choice of the field's direction was completely arbitrary, but this convention prevents a lot of confusion when scientists are discussing magnetic fields. The steps described below will model a magnetic field's direction.

D Tape the five compasses to a piece of cardboard so that they are equally spaced around a hole just large enough to allow the bar magnet to pass through (see above left). Avoid taping them too close together so that they don't affect each other.

E Holding the cardboard horizontally, allow the compass needles to come to rest. Slowly pass the magnet vertically through the hole in the cardboard. Note the pole that is entering first and the orientation of the compass needles.

F Have your lab partner finish pulling the magnet through the hole from underneath the cardboard. Note the direction of the compass needles as this process occurs.

G Fill in Diagram A to show the orientation of the compass needles when the north pole of the magnet is in the hole. After drawing the compass needles, draw in the field lines associated with the end of the magnet and indicate the direction of the field. Remember that *the field direction is the same direction that the north end of the compass needle is pointing.*

Diagram A

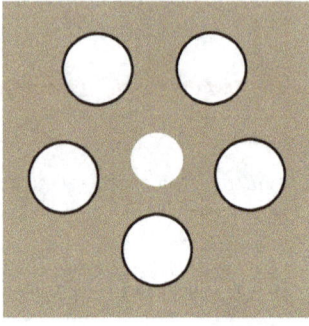

H Repeat Step G with the south pole of the magnet in the hole and draw your results on Diagram B.

3. In what direction does the field point at the north end of the magnet?

_____

4. In what direction does the field point at the south end?

_____

Diagram B

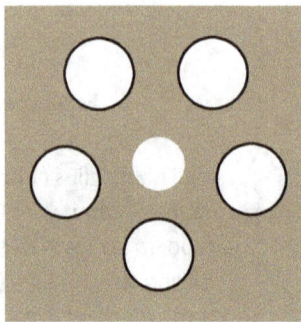

Name _____

Modeling Field Lines

As previously mentioned, scientists use lines to represent the shape of a magnetic field. Remember that field lines are a way to model the field visually. Magnetic particles tend to line up parallel to the forces in a field. Your eyes and brain interpret this arrangement of particles as lines.

I Lay one bar magnet flat on the table. Take the two furring strips and arrange them on each side of the magnet to support the sheet of cardstock, laying the cardstock over them.

J Carefully sprinkle the iron filings onto the cardstock around and on top of the magnet. Do not pour the filings in a pile. As more particles are trapped by the field, tap the cardstock with a pencil to help the particles become oriented with the field. Use enough filings so that all parts of the field can be seen.

Diagram C

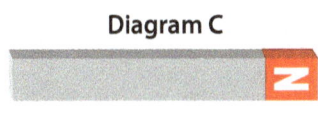

K Using Diagram C, sketch the shape and direction of the lines around the bar magnet.

L When you are finished, lift the cardstock straight up, bend it to form a trough, and pour the filings back into the container.

Diagram D

 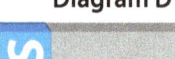

M Place two bar magnets so that their opposite poles are in line with each other but separated by about 3 cm.

N Arrange the furring strips and center the sheet of cardstock over the gap between the ends of the magnets.

O Repeat Step J and draw the shape and direction of the lines around the magnets in Diagram D.

Diagram E

P Return the iron filings to the container. Rotate one of the magnets so that the two north poles are now facing each other. Replace the cardstock.

Q Repeat Step J and sketch the shape and direction of the field on Diagram E.

R Return the filings to the container and clean up any stray particles.

5. On the basis of your observations in Questions 1–2 and your drawings in Diagrams C–E, describe the density of the field lines at the end of a magnet compared with its middle.

_____

_____

_____

6. What generalization can you make regarding line density and field strength?
   _____
   _____
   _____

7. Describe the shape and behavior of field lines between opposite poles of two magnets.
   _____

8. Describe the shape and behavior of field lines between like poles of two magnets.
   _____

9. State two ways that the drawing of field lines on a flat piece of paper imperfectly models the actual magnetic field.
   _____
   _____
   _____
   _____

GOING FURTHER

Engineers have used the repelling power of magnets in transportation. A maglev train uses the repelling power of magnets to raise the train above its track before it begins moving.

10. Why would raising the train above its track be advantageous?
    _____
    _____
    _____

11. Would a maglev train be able to move without any friction?
    _____
    _____

12. How might reducing friction when moving a train be an example of wise stewardship?
    _____
    _____
    _____
    _____
    _____

CHAPTER 20: MAGNETISM

LAB 20B
MIGHTY MAGNETS

Inquiring into Electromagnets

Electromagnets have many uses. Large electromagnets move iron and steel scrap around. Electromagnets hold some fire-rated doors open in schools and other public buildings unless the fire alarm goes off. One of the most exciting technologies using electromagnets to emerge in the past forty years is the MRI, which uses powerful electromagnets to obtain images of the internal structures of the body that cannot be observed using x-rays. MRIs are also safer than x-rays.

As you learned in your textbook, an electromagnet consists of a ferromagnetic bar placed inside a solenoid with current running through it. But not all electromagnets are equally strong. Some can pick up much more weight than others. In this activity, you will create a method for testing the strength of electromagnets and then use it to determine how changing different components of an electromagnet affects its strength.

QUESTIONS

How can I build a stronger electromagnet?

- How can I construct an electromagnet?
- What variables affect the strength of an electromagnet?

PROCEDURE

Building a Base-Model Electromagnet

A Build a base-model electromagnet by wrapping the wire around the bolt 50 times, leaving at least 20 cm free at each end. Tape the wire to the bolt to keep it from unwrapping.

B Connect the free ends of the wire to the battery holder. When you are ready to test the electromagnet, place the two AA batteries in the battery holder. Leave the electromagnet connected only long enough to test it. If you notice any part of the electromagnet becoming hot, immediately disconnect one end of the wire and notify your teacher.

Equipment
22-gauge insulated wire, 2 m
3/8 in. steel bolt, 6 in. long
masking tape
battery holder
AA batteries (2)

Mighty Magnets 211

Planning/Writing Scientific Questions

C Make a list of the parts of an electromagnet.

D Make a list of variables that may affect how well an electromagnet works.

E Write questions that will help guide your process of testing the way that changes in the variables in Step D affect an electromagnet's performance.

Designing Scientific Investigations

F Write procedures that will allow you to test each part of an electromagnet. Think about steps that will allow you to test changes to each part separately. Decide on a way to test the strength of the magnet and choose the number of trials that you should run.

G Create a list of materials that you will need to complete your testing procedures.

H Have your teacher approve your procedures and materials list.

Conducting Scientific Investigations

I Obtain materials from your teacher.

J Follow your procedures to test the way that different variables affect the strength of an electromagnet.

1. Of the parts of the electromagnet that you tested, which had the most influence on its strength?

   _____

   _____

2. How did you change the part(s) in Question 1 to increase the electromagnet's strength?

   _____

   _____

3. Which component had the least influence on the strength of the electromagnet?

   _____

   _____

4. Choose one of the variables that you tested. If you were to increase that variable 1.5 times, how much would the strength of the magnet increase? Make a testable prediction.

   _____

   _____

5. Build an electromagnet according to your prediction and test it. Did the results support your prediction? Explain.

   _____

   _____

CHAPTER 21: ELECTROMAGNETIC ENERGY

LAB 21A

LIGHT LIMIT

Investigating Changes in Light over Distance

Have you ever taken a walk on a really dark night? Perhaps you were camping somewhere far away from city lights and the moon hadn't risen yet. If you were smart, you took a flashlight with you so that you could find your way around. You quickly learned something about the nature of light: while the flashlight easily illuminated the ground immediately in front of you, it was much less effective at lighting things farther away.

As you've also learned from your textbook, electromagnetic energy such as light travels as photons. These particles originate at the light's source and move outward through space. Normally they travel in straight lines unless they encounter something that changes their direction.

The amount of light energy that leaves a light source every second is called the *luminous flux* and is measured in the SI unit of *lumens* (lm). You may be familiar with this unit if you have ever purchased light bulbs. Common lightbulbs used in homes emit from 300 lm for a 40 W bulb to 1800 lm for a 100 W bulb. When we talk about how effective a light source is at illuminating an object, we are interested in illuminance, which is the amount of lumens that land on each square meter of surface. The SI unit for illuminance is the *lux* (lx).

In this lab activity, you will investigate how illuminance changes with distance and why it changes the way that it does.

QUESTIONS

Why does a flashlight have a limited useful range?

- Does a light source's effectiveness vary in a predictable way?
- Can we predict the effectiveness of a light source over different distances?

Light Limit 213

Equipment
Labdisc Gensci
computer
masking tape
meter stick
LED flashlight
construction paper

PROCEDURE

Changing Illuminance with Distance

Light's effectiveness decreases with distance, and scientists are always interested in making accurate predictions. Does the change in light's effectiveness follow a predictable pattern? Let's look at exactly how illuminance changes as the distance between a light source and an illuminated object changes.

A Tape a meter stick to the surface of your laboratory table.

If you have not used the Labdisc Gensci before, refer to Appendix F for more detailed instructions.

B Set the Labdisc Gensci to collect light meter data.

C Set the **Sampling rate** to **10/sec**.

D Set the **Number of samples** to **100**.

E Place the Labdisc Gensci on the laboratory table so that the light sensor is aligned with the 0 cm mark of the meter stick and facing the length of the meter stick.

F Reduce the lighting in the room as much as possible.

G Turn on the flashlight and place it on top of the meter stick, facing the light sensor, so that the illuminated end is on the 20 cm mark. The light will thus be 20 cm from the light sensor. Your setup should look similar to the image below.

H Notice the value for illuminance on the Labdisc Gensci. Start the data collection, which will run for 10 seconds.

1. State a hypothesis about what will happen to the illuminance if the flashlight is moved to 40 cm away, that is, twice the distance.

   _____
   _____
   _____

I Repeat Steps G and H with the flashlight set at 40 cm.

2. Did the illuminance change as you expected? Explain.

   _____
   _____
   _____

J Repeat Steps G and H with the flashlight set at 60 cm, 80 cm, and 100 cm.

K Connect the Labdisc Gensci to a computer and download the 20 cm data set. Use the **Statistics** tool to get the average illuminance. Record this value in Table 1.

L Repeat Step K for the 40 cm, 60 cm, 80 cm, and 100 cm data.

Modeling the Phenomenon

M Create a scatterplot of your data for illuminance (*y*-axis) and distance (*x*-axis) values on Graph 1. Make sure that you draw a smooth curve that represents all the data. This is your curve of best fit.

3. Does the illuminance change linearly with distance? Explain.

   _____

   _____

4. Does this model explain why a flashlight's ability to illuminate distant objects isn't very good? Explain.

   _____

   _____

   _____

Predicting with Your Model

N Either graphically or algebraically, use your model to predict the illuminance when the flashlight is placed at the 50 cm mark. Record your prediction in Table 2.

O Either graphically or algebraically, use your model to predict the distance at which the illuminance will be 800 lx. Record your prediction in Table 2.

P Use your experimental setup to test your predictions. Record the actual illuminance and distance in Table 2.

Q Use the percent error formula to calculate the percent error for your two predictions and record these values in Table 2.

$$\%_{error} = \left(\frac{\text{predicted value} - \text{actual value}}{\text{actual value}}\right) 100\%$$

5. Is your model workable—does it explain observations and make accurate predictions? Explain.

_____

_____

Why Does Light Behave This Way?

Does light become less effective over a distance because the photons are losing energy as they move away from the light source? In other words, do they "get tired"?

The image on the left shows a light source radiating photons in every direction. These photons fill an imaginary spherical region around the lamp as they move out into space. The effectiveness of a light source at illuminating an object depends on how many lumens reach the object. In the case of the 100 W bulb mentioned earlier, 1800 lm of light energy are spread across the surface area of the imaginary sphere surrounding the light. As the distance between the light source and the illuminated object increases, the sphere of light energy gets larger and larger. Those 1800 lm of light get spread over a bigger area. Let's see how this works.

R Hang the construction paper on a wall.

S Turn on the flashlight and hold it so that the lamp is approximately 1 m from the paper. Adjust your distance if the circle of illumination is not entirely on the paper. Orient the flashlight so that the illuminated area is a circle. Record your distance in Table 3.

T Have a second person trace the outline of the illuminated area of the paper.

U Move forward toward the paper to half of your original distance.

6. Did you change the amount of energy leaving the flashlight as you walked toward the construction paper?

_____

V Again have a second person trace the outline of the illuminated area of the paper. Record your distance in Table 3.

7. When you cut the distance in half, did the brightness of the light on the wall change? Explain.

_____

_____

W Measure the radius of each circle. Record these radii in Table 3.

X Calculate the area of each circle and record these areas in Table 3.

Y Divide the smaller area by the larger area and record this ratio in Table 3.

8. When you cut the distance in half, what happened to the area of illumination?

   _____

9. How can you explain the brighter illumination when you moved closer to the paper?

   _____

   _____

   _____

10. Does the decrease of illumination with distance from a light source appear to be caused by "tired" photons? Explain.

    _____

    _____

    _____

Going Further

What you have observed in this lab activity is called an inverse square law. This is similar to the inverse relationships that you learned about in math class.

Inverse Function

$$y = \frac{k}{x}$$

Inverse Square Function

$$y = \frac{k}{x^2}$$

In an inverse function, as x changes, y changes by the inverse of the same factor. For example, in Ohm's law (see Lab 19A),

$$I = \frac{V}{R}.$$

Current and resistance are inversely related. This means that if resistance is doubled, current will be cut in half. The inverse square law works like this except that the variable in the denominator is squared. The formula for illuminance of the illuminated construction paper is

$$I = \frac{P_{light}}{\pi r^2},$$

where I is the illuminance in lumens per square meter, P_{light} is the luminous flux in lumens, and r is the radius of the circle in meters.

11. Your flashlight emits 300 lm that land on a circle with a radius of 0.9772 m. Calculate the illuminance of the circle.

12. You move the flashlight twice as far away, so the radius is twice as much—1.954 m. What is the illuminance now?

13. What fraction of the original is the reduced illuminance?
    _____

14. Thinking that the illuminance is too low, you decide to fix the problem by using a flashlight that emits 600 lm. What is the illuminance of this flashlight on a circle with a radius of 1.954 m?

15. Did the more powerful flashlight entirely offset the decrease of illuminance caused by the increased distance? Explain.
    _____
    _____

16. Which has a bigger impact on illuminance, changing distance or changing luminous flux? Explain.
    _____
    _____

Table 1

DISTANCE (cm)	AVERAGE ILLUMINANCE (lx)
20	
40	
60	
80	
100	

Table 2

DISTANCE (cm)	PREDICTED ILLUMINANCE (lx)	ACTUAL ILLUMINANCE (lx)	% ERROR
50			

ILLUMINANCE (lx)	PREDICTED DISTANCE (cm)	ACTUAL DISTANCE (cm)	% ERROR
800			

Table 3

DISTANCE (m)	RADIUS (cm)	AREA (cm^2)	RATIO

Light Limit

CHAPTER 21: ELECTROMAGNETIC ENERGY

LAB 21B
DRIVEN TO DIFFRACTION

Investigating the Bending of Light

After Noah left the ark, God placed the rainbow in the sky as a sign that He would not destroy the world by flood again. Today people love seeing the stunning display of light in a rainbow.

Rainbows have been observed and enjoyed for thousands of years, but it wasn't until Isaac Newton's day that people began to understand the science behind them. Newton was the first to realize that white light is the combination of all the different colors of light. Using a glass prism, he *dispersed* (separated) white light into its component colors. Scientists later explained that this phenomenon occurred because the speed of light differed depending on its color. Different colors, with their different speeds, bend at different angles, resulting in a spectrum of color—a rainbow.

Later scientists discovered that there was much more to light than meets the eye. They discovered that light also diffracts as it moves around an obstacle or through an opening. The degree of bending is related to the wavelength of light. In other words, different colors of light will diffract at different angles. In this lab activity, you will investigate the bending of light through a diffraction grating. You will model the relationship between the light's wavelength and the degree to which it is bent. You will also attempt to experimentally determine the number of lines per millimeter in the diffraction grating.

QUESTIONS

Why does light form a rainbow?

- How can we make a rainbow?
- What determines how much light bends?

PROCEDURE

Data Collection

Equipment
flashlight, single LED
meter stick
diffraction grating
computer paper (3 pcs.)

A Tape the three pieces of paper together along their short sides to form a long piece of paper that is roughly 21.6 cm × 83.8 cm.

B Lay the paper on the laboratory table with the meter stick laying along the short edge of the paper. Arrange the paper and meter stick so that the 100 cm end of the meter stick is close to an edge of the table (see below).

C Set the flashlight so that its bulb is aligned with the short edge of the paper and the light shines along the meter stick.

The diffraction grating that you are using in this activity is a clear piece of plastic with tiny, evenly spaced lines etched into it. The light bends as it moves through the openings created by those lines. Since this diffraction grating is clear and light is transmitted through it, we call it a *transmissive diffraction grating*. Other types of diffraction gratings are considered reflective because they diffract the light that is reflected from them. If you have ever noticed the rainbow pattern reflected off the surface of a CD, you have noticed the diffraction caused by the tracks of data on the CD.

D Dim the lights in the room as much as possible.

E Look through the diffraction grating as you hold it 1 m from the flashlight. To the side of the light you should see a spectrum of color. (If you can't, you may need to rotate the diffraction grating 90°.) Adjust the angle of the diffraction grating so that the spectrum is on top of the paper.

F Direct a lab partner to mark on the paper the center of the blue, green, yellow, and red regions of the spectrum.

If you look farther to the side of the spectrum that you have been working with, you should notice additional sets of colors. We are working only with this first and brightest spectrum. Physicists identify this spectrum as $m = 1$.

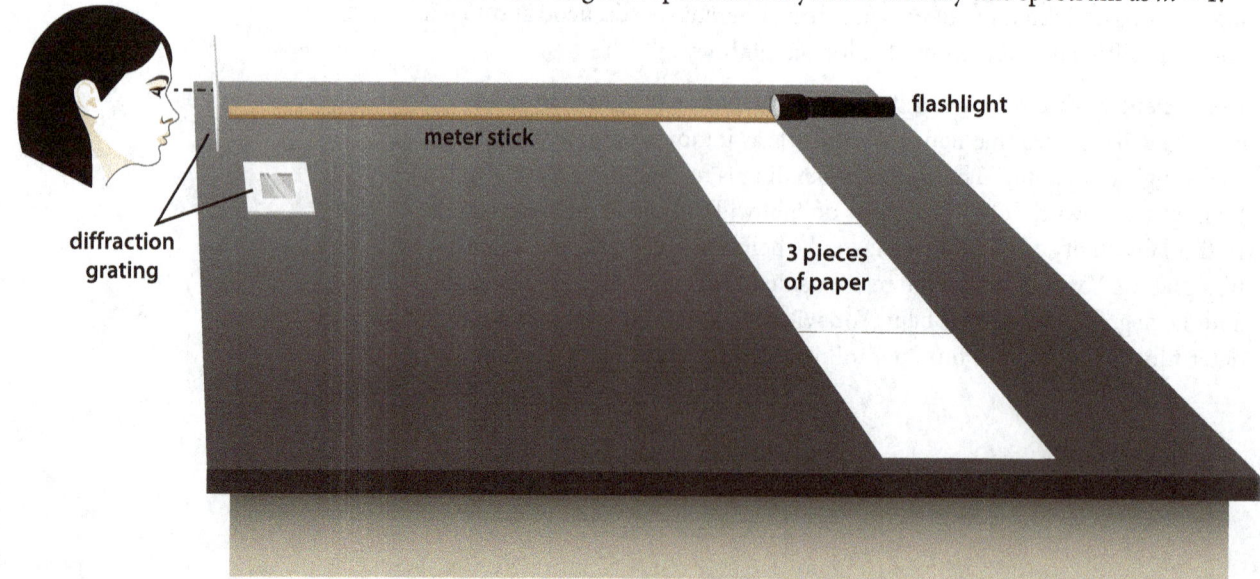

1. Which of the four colors diffracts the most? Which diffracts the least?

   _____

G. Measure the distance from the edge of the paper where the flashlight was to the marks for each of the four colors. Record these values in Table 1.

Developing and Testing Models

2. On the basis of your data, what is the general trend of bending in relationship to wavelength?

   _____

   _____

H. Draw a scatterplot of the diffracted distance versus wavelength in the graphing area. Include a curve of best fit.

I. Using your graph, predict the diffracted distance for the center of the violet and orange portions of the visible light spectrum. Record your predictions in Table 2.

J. Use the experimental setup to measure the distance to the center of the violet and orange regions of the spectrum. Record the distances in Table 2.

K. Calculate the percent difference between your predictions and the measured values. Record these values in Table 2.

Grating Spacing

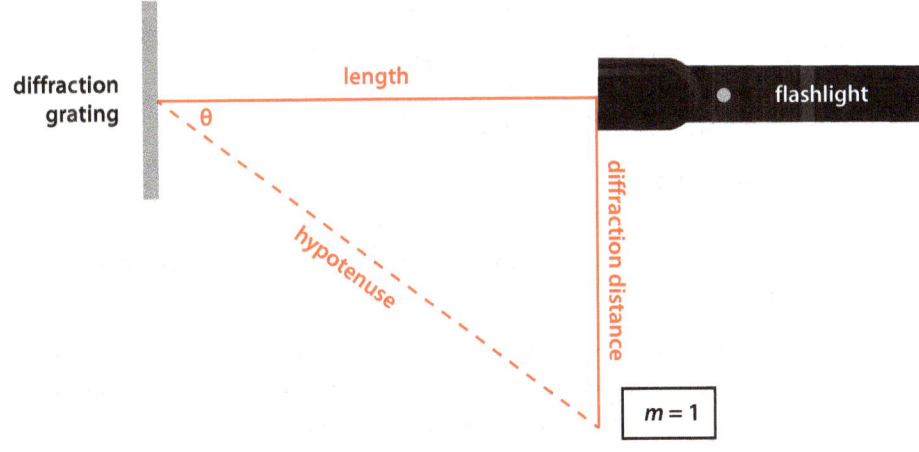

As you have seen in this activity, the amount of bending is related to the wavelength of light. Another factor that affects the amount of diffraction is how closely the lines in the diffraction grating are. For this lab activity, you will use the formula below to solve for d, the grating spacing. This formula contains the variables that you measured in the activity.

$$d = \frac{\lambda \sqrt{distance^2 + length^2}}{distance}$$

Driven to Diffraction

L Use the formula and your data for blue and green light to calculate the diffraction grating spacing—in both mm/line and lines/mm. Record your values in Table 1. (*Note:* All your measurements will need to be in the same units; millimeters will probably work best.) Notice the example calculation for the following sample data (see below the table).

Sample Data

	LENGTH (cm)		100	
COLOR	CENTER WAVELENGTH (nm)	MEASURED DISTANCE (cm)	SPACING, d (mm/LINE)	LINES/mm
red	683	37.20	1.96×10^{-3}	510

Unit Conversions:

$$\lambda = 683 \text{ nm} \left(\frac{1 \text{ m}}{1 \times 10^9 \text{ nm}}\right)\left(\frac{1 \times 10^3 \text{ mm}}{1 \text{ m}}\right) = 6.83 \times 10^{-4} \text{ mm}$$

$$\text{distance} = 37.20 \text{ cm} \left(\frac{1}{1 \times 10^2 \text{ cm}}\right)\left(\frac{1 \times 10^3 \text{ mm}}{1 \text{ m}}\right) = 3.720 \times 10^2 \text{ mm}$$

Separation Distance Calculation:

$$d = \frac{6.83 \times 10^{-4} \text{ mm} \sqrt{(3.720 \times 10^2 \text{ mm})^2 + (1.000 \times 10^3 \text{ mm})^2}}{3.720 \times 10^2 \text{ mm}}$$

$$= \frac{6.83 \times 10^{-4} \text{ mm} \sqrt{1.38384 \times 10^5 \text{ mm}^2 + 1.000 \times 10^6 \text{ mm}^2}}{3.720 \times 10^2 \text{ mm}}$$

$$= \frac{6.83 \times 10^{-4} \text{ mm} \sqrt{1.138384 \times 10^6 \text{ mm}^2}}{3.720 \times 10^2 \text{ mm}}$$

$$= \frac{(6.83 \times 10^{-4} \text{ mm})(1.066950795 \times 10^3 \text{ mm})}{3.720 \times 10^2 \text{ mm}}$$

$$= 1.9589446 \times 10^{-3} \text{ mm/line}$$

$$\frac{1}{d} = 510 \text{ lines/mm}$$

M Obtain the actual value for the diffraction grating spacing (in lines per mm) from your teacher.

3. On the basis of your predicted diffraction and your diffraction grating spacing calculation, state a claim about how workable your model is. Support your claim with evidence from your data.

_____

_____

GOING FURTHER

This fascinating interaction of matter and electromagnetic energy is studied in the field of *spectrometry*. Scientists use spectrometry in many applications, such as testing for lead in blood. Lead poisoning is a serious health risk, especially for children. Lead builds up over time, even when the exposure amounts are relatively low. Lead poisoning can cause both mental and physical developmental problems. This is why the federal government has taken steps to limit exposure to lead, including a ban on leaded gasolines and prohibitions regarding lead-based paints.

4. Why do you think that children, especially those six years old and younger, are most affected by lead poisoning?

   _____
   _____
   _____
   _____
   _____
   _____
   _____
   _____
   _____
   _____

5. People don't come in contact with gasoline often, so why would lead in gasoline be a concern?

   _____
   _____
   _____

6. Many older houses in the United States were painted with lead-based paints. While the paint is undisturbed, it is relatively safe, but when it is disturbed, the lead can be ingested. When would lead-based paints be the greatest threat?

   _____
   _____
   _____

Table 1

COLOR	LENGTH (cm) CENTER WAVELENGTH (nm)	MEASURED DISTANCE (cm)	100 SPACING, d (mm/LINE)	LINES/mm
blue	467.5			
green	532.5			
yellow	577.5			
red	682.5			

Table 2

COLOR	WAVELENGTH (nm)	PREDICTED DISTANCE (cm)	MEASURED DISTANCE (cm)	PERCENT DIFFERENCE
violet	415			
orange	608			

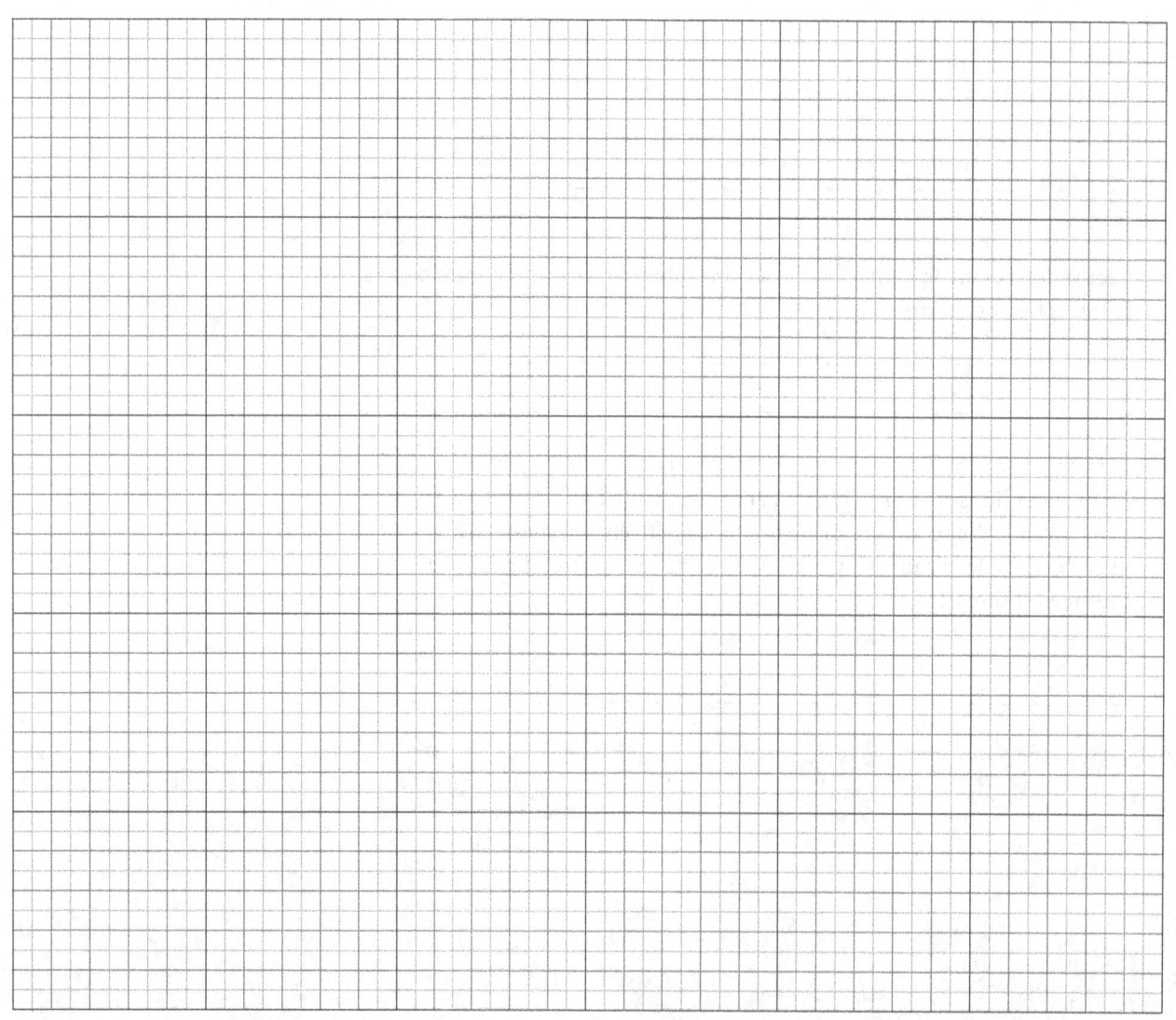

CHAPTER 22: LIGHT AND OPTICS

Name
Date

LAB 22A
UPON REFLECTION

Investigating Mirrors and Virtual Images

Many carnivals and fairs have a hall of mirrors. These amusements feature mazes or obstacle courses made of mirrors. Many of these mirrors are curved, distorting images to startling effect.

Mirrors around your house are more likely to be plane mirrors—they are flat. It's this type of mirror that you will use in this activity to investigate virtual images and test the law of reflection.

QUESTIONS

How does a mirror work?

- What is a virtual image?
- Is the law of reflection really true?

PROCEDURE

Virtual Images

A Using the ruler, draw a line across a piece of paper near one end and label the line *l*. Tape the paper to the piece of cardboard.

B Use two pieces of masking tape to secure the mirror to the mirror support. Position the edge of the mirror's reflective surface on *l* as shown below.

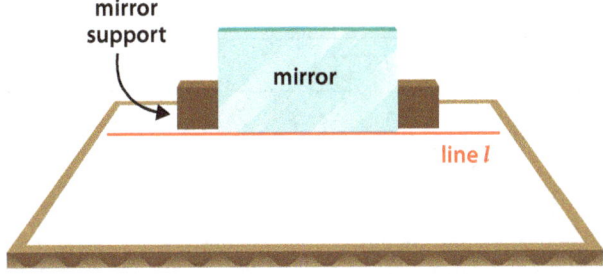

Equipment

metric ruler
plane mirror
mirror support
protractor
white paper
masking tape
cardboard sheet, 30 × 30 cm
straight pins (5)
goggles

Upon Reflection 227

C Stick a pin into the paper about 3 cm in front of the middle of the mirror. Write an O (for object) by the base of the pin. We'll call this point O.

D Place another pin about 7 cm in front of and 7 cm to the left of O. Label this point A.

E With one eye closed, bring down your open eye to the level of the cardboard and sight a line that passes from the base of the pin at A to the base of the reflection of the pin at O in the mirror. Mark this line of sight by placing another pin into the paper at a point between A and the image of O. Label this point B (see below).

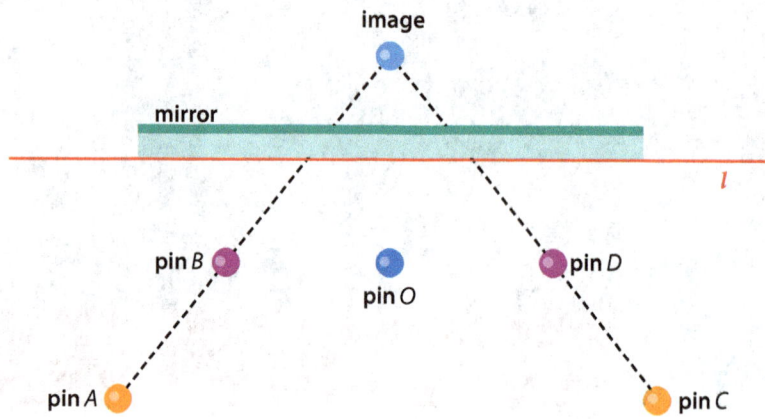

F Place another pin about 7 cm in front of and 7 cm to the right of O. Label it C.

G Repeat Step E using C. Label the new point D.

H Remove the mirror and all pins from the paper. Draw a line through A and B. Extend the line 8–10 cm past l. Draw a line through C and D, extending it until it intersects line $\overleftrightarrow{AB}$. Draw a dot at the intersection. This dot is the location of the virtual image. Label this point V.

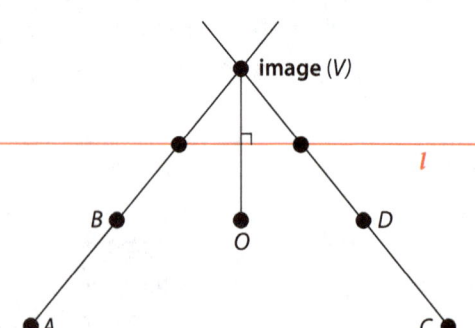

I Draw a line connecting O and V. This line is the normal line and should be perpendicular to l (see left).

J Measure the distances from O to l and from V to l. Record these measurements in Table 1.

1. Compare the distances that you measured in Step J. What can you conclude from your comparison?

_____

_____

_____

2. How can you be sure that V is the position of O's virtual image?

3. How do you know that the image in the mirror is virtual rather than real?

The Law of Reflection

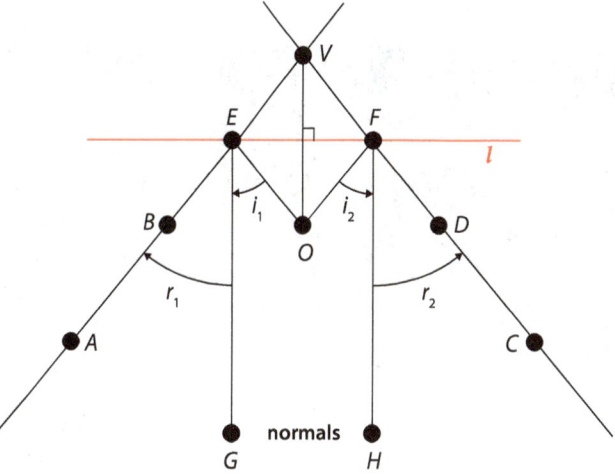

K Label the points where $\overleftrightarrow{AB}$ and $\overleftrightarrow{CD}$ intersect l as E and F, respectively.

L Draw line segments from O to E and F (right).

M Use a protractor to draw two lines perpendicular to l at E and F. Label points G and H as shown.

N Measure to the nearest 1° the angles of incidence (i_1 and i_2) and the angles of reflection (r_1 and r_2). For example, the first incident angle, $\angle i_1$, is $\angle OEG$. Its reflected angle, $\angle r_1$, is $\angle AEG$. Extending the lines with a pencil and ruler makes it easier to measure the angles. Record your results in Table 1.

4. Compare each angle of incidence to its corresponding angle of reflection. Does your comparison seem to confirm the law of reflection? Explain.

5. If the values for the pairs of angles that you measured are not identical, is the law of reflection incorrect? Explain.

6. List some potential sources of error in this exercise. Be specific.

7. How could you improve your confidence in your results for this experiment?

   _____
   _____
   _____

8. Do you think that the law of reflection would apply to a convex mirror? Explain.

   _____
   _____
   _____

GOING FURTHER

A virtual image in a mirror appears to be the same distance behind the mirror as the reflected object is in front of the mirror. Because of this fact, large mirrors can make a room look larger than it really is. This fact was taken to the extreme in the Hall of Mirrors at the Palace of Versailles in France. The hall was not a carnival attraction. In fact, King Louis XIV sometimes received distinguished foreign ambassadors there. But the mirrors do distort reality. The Hall of Mirrors is 73 m long but only 10.5 m wide. But it looks much bigger than it actually is because it contains 357 mirrors.

9. A person stands at one side of the Hall of Mirrors and looks at a mirror on the other side, 10 m away. He sees the reflection of a pillar that is 10 m from the mirror. How far away from the person does the reflection of the pillar appear to be?

   _____

10. Besides making a room look larger, what other interior decorating purposes can a mirror be used for?

    _____
    _____
    _____

Table 1

DISTANCE O TO I (cm)		
DISTANCE V TO I (cm)		
$\angle i_1$ (°)		
$\angle r_1$ (°)		
$\angle i_2$ (°)		
$\angle r_2$ (°)		

CHAPTER 22: LIGHT AND OPTICS

LAB 22B
THROUGH THE LENS OF THE BEHOLDER

Exploring Lenses

Do you wear glasses or contacts? If not, you almost certainly know someone who does. These corrective lenses help to correct vision problems by focusing light rays. Eyeglasses were first invented in the 1200s, but the science of optics didn't really advance until the 1600s. Eyeglasses did not become widely available to the general public until the 1800s.

We'll return to corrective lenses in a little bit, but first you are going to explore the properties of a converging lens. Before you begin, let's review some basic lens terminology.

As you learned in your textbook, a lens's *optical axis* is an imaginary line perpendicular to the surface of the lens that passes through the center of the lens. The *lens plane* is an imaginary line parallel to the surface of the lens that passes through the center of the lens. The point where these two lines cross is the lens's *optical center* (see next page).

QUESTIONS

How is optics related to vision?

- How does a convex lens affect light rays?
- How is a lens's focal length related to the type of image that it produces?
- How is the object-lens distance related to the type of image that the lens produces?

PROCEDURE

Determining Lens Focal Length

Equipment
converging lens
lens holder
screen
screen holder
calculator
meter sticks (2)
meter stick supports (2)
light source
light holder
goggles

You learned in your textbook that parallel light rays are focused by a converging lens at the lens's focal point. But light rays will be parallel only if they come from an object infinitely far away. Of course, no such object exists. Real objects are a finite distance away, so their light rays are not parallel. Therefore, these objects will be in focus at locations other than the focal point. The distance at which the image of a point of an object is in focus is called the *image distance* (see below).

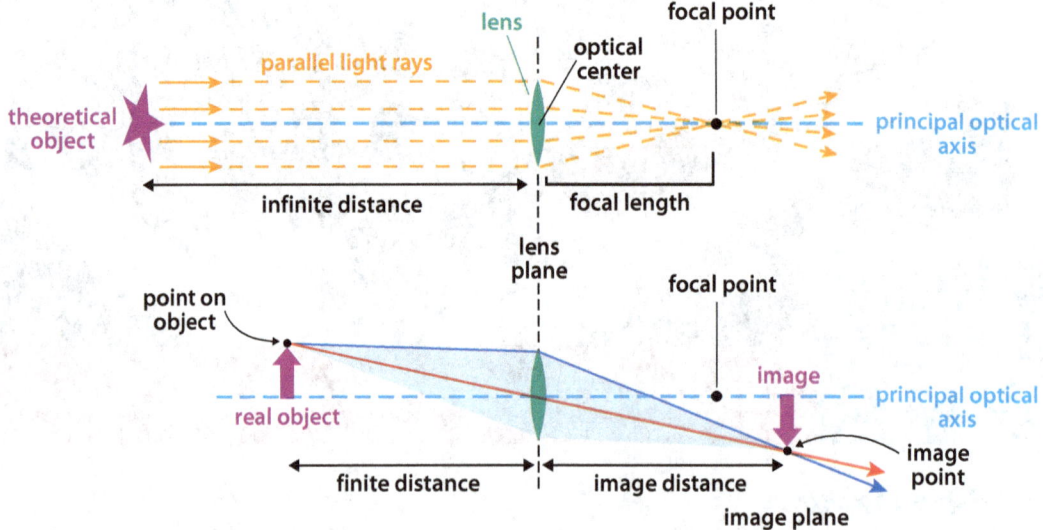

But if a lens's image distance is equal to its focal length only when the object is infinitely far away, how can you find the lens's focal length? For all practical purposes, a lens will focus light rays from a faraway object very near the focal point. In this part of the activity, you will use this fact to calculate the focal length of a lens.

A Insert the lens and the screen into their holders. Set the meter stick into its support legs and clip the screen on top of the meter stick at the 50 cm mark. Clip the lens holder onto the meter stick between the screen and the 0 cm mark (as shown below).

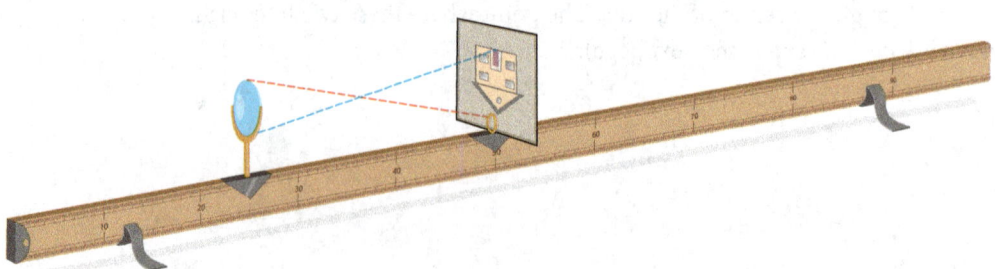

B Point the 0 cm end of the meter stick toward a distant object more than 10 m away, preferably a brightly lit object outside the building. Slide the lens along the meter stick until the image on the screen is in sharp focus.

232   Lab 22B

C Note the positions of the screen and the lens on the meter stick and record them in Table 1. Subtract the lens position from the screen position to obtain the lens's focal length. Record it in Table 1.

D Slide the lens support in either direction until the image is out of focus. Then repeat Steps B and C using other distant objects. Record the data in Table 1. If possible, different people should perform each trial.

E Calculate the average lens focal length using the three values you found and record it in Table 1.

1. Describe the image orientation on the screen compared with the original object.
   _____

2. How did the image size compare with the actual object size?
   _____

3. Why is using a very distant object for determining the focal length of a lens better than using a nearer object?
   _____
   _____
   _____

Real Images

Now that we know the lens's focal length, we can examine the lens's behavior under a variety of conditions. In this section, you are going to place an object at distances from the lens that are specific multiples of its focal length (f) to see what kind of image the lens forms.

F Unclip the screen from the meter stick. Slide the lens to the 50 cm mark.

G Insert the light source into the light holder and clip the holder onto the meter stick at a distance 2.5 × f from the lens.

H Clip the screen onto the meter stick on the other side of the lens. Turn on the light.

I Slide the screen along the meter stick until the image on the screen is in sharp focus.

J Measure the distance between the lens and the screen. This distance is d_i, the *image distance*. Record the image distance in Table 2.

K Measure the distance between the lens and the light. This distance is d_o, the *object distance*. Record the object distance in Table 2.

L Divide d_i by d_o to calculate the magnification. Record this ratio in Table 2.

Through the Lens of the Beholder

M   Note whether the image is larger or smaller than the object, whether it is upright or inverted, and whether it is a real or virtual image. Record your observations in Table 2.

4. How do you know whether the image is real or virtual?
   _____
   _____

N   Move the light to a distance of $2 \times f$ from the lens. Repeat Steps I–M. Record all data in Table 2.

O   Move the light to a distance of $1.5 \times f$ from the lens. Repeat Steps I–M. Record all data in Table 2.

5. What do you notice about the magnification ratio and the size of the image compared with the object?
   _____
   _____
   _____

6. What do you notice about the relationship between the object distance, the focal length, and the magnification of the image?
   _____
   _____
   _____

P   Slide the light slowly from the $1.5 \times f$ position through the $1 \times f$ position. Move the screen as necessary to keep the image in focus.

7. What appears to happen to the image as you approach the lens's focal point?
   _____
   _____

Virtual Images

Q   Move the light and the lens so that they are separated by a distance of $0.5 \times f$. They should be near the 0 cm end of the meter stick. Slide the screen to obtain the clearest image possible.

8. Were you able to form a focused image on the screen? Explain.
   _____

R   Remove the screen from the meter stick.

S   Close one eye and slowly move your head toward the lens while looking at the light.

T Note whether the light source as seen through the lens appears larger or smaller than the source itself, whether it is upright or inverted, and whether it is a real or virtual image. Record your observations in Table 2.

9. On the basis of your observations from this lab activity, what do you think are two important differences between real and virtual images formed by a converging lens?

_____
_____
_____

10. Water droplets, such as the ones shown at right, often form natural lenses. Is the image formed by the droplet lens in this photo real or virtual? Explain.

_____
_____

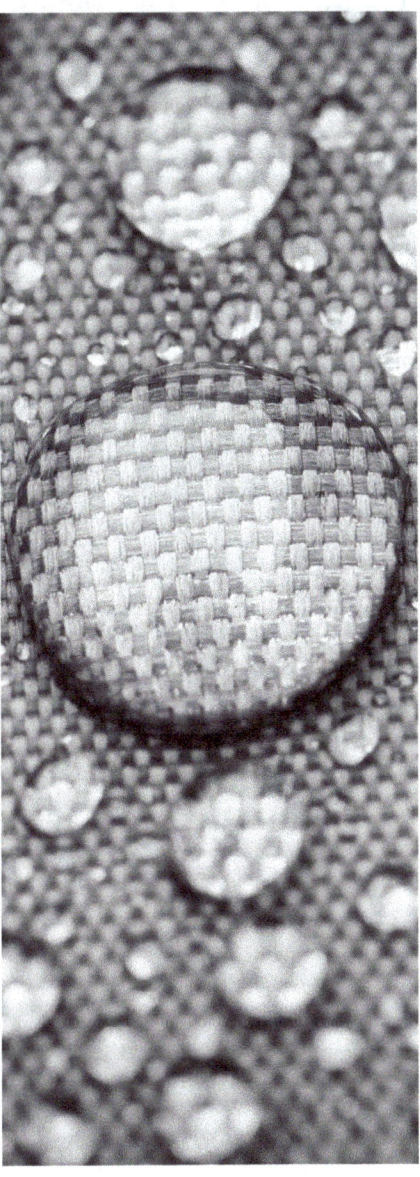

GOING FURTHER

Your eye is a converging optical system consisting of several structures that refract light. Most of the refraction is accomplished by the outer layer of the eye—the cornea. The lens performs much of the rest of the refraction. The aqueous and vitreous humors also refract light to some degree. This refraction focuses images on the retina, the inner layer of the eye. Structures called photoreceptors transform the light rays into neural signals that travel along the optic nerve to the brain.

11. The lens of the eye changes shape, changing its focal length. How is this different from the lens that you used in this lab activity?

_____
_____
_____

12. How is the characteristic described in your answer to Question 11 useful for vision?

_____
_____

In many people, the cornea becomes elongated, resulting in the image being focused in front of the retina. Images of objects close to the person are focused on the retina, so that these images are clear. This condition is called *myopia*. In many cases, corrective lenses can correct this issue.

13. How is manufacturing corrective lenses an example of using optics to fulfill the command to love our neighbor?

_____
_____
_____

Table 1

TRIAL	SCREEN POSITION (cm)	LENS POSITION (cm)	FOCAL LENGTH (cm)
1			
2			
3			
Average			

Table 2

OBJECT DISTANCE	IMAGE DISTANCE (cm)	OBJECT DISTANCE (cm)	MAGNIFICATION (d_i/d_o)	IMAGE SIZE	IMAGE ORIENTATION	IMAGE TYPE
$2.5 \times f$						
$2.0 \times f$						
$1.5 \times f$						
$0.5 \times f$						

APPENDIX A
LABORATORY AND FIRST AID RULES

Laboratory Rules

1. Never perform an unauthorized experiment or change any assigned experiment without your teacher's permission.
2. Avoid playful, distracting, or boisterous behavior.
3. Never work alone. Students must not conduct lab activities without supervision.
4. Work at your own lab station.
5. Always work in a well-ventilated area. Use the fume hood when working with toxic vapors. Never put your head in the vent hood.
6. Always wear safety goggles when working with chemicals, glassware, projectiles, and other materials or objects that are potentially hazardous to the eyes.
7. Wear protective clothing and gloves when working with corrosive or staining chemicals.
8. While working in the laboratory, tie back long hair and avoid wearing loose clothing such as scarves or ties.
9. Never taste any chemical, eat or drink out of laboratory glassware, or eat or drink in the laboratory.
10. Always use the appropriate instruments for cutting and handle them carefully. Always cut away from yourself.
11. When handling live organisms, follow instructions and do not cause them undue harm or discomfort.
12. Thoroughly wash your hands with soap after handling any live organisms, cultures containing organisms, or chemicals.
13. To smell a substance, gently fan its vapor toward you.
14. Never leave a flame or heat source unattended. Keep combustible materials away from heat sources.
15. When diluting acid solutions, always add the acid to water slowly. **Never add water to an acid!**
16. When heating a test tube, point the open end away from you and others. **Never heat a closed or stoppered container!**
17. Dispose of waste as instructed by your teacher.
18. Do not return unused chemicals to a container. Dispose of them properly.
19. Notify the teacher of any injuries, spills, or breakages.
20. Know the locations of the fire extinguisher, safety shower, eyewash station, fire blanket, first-aid kit, and Safety Data Sheets.

APPENDIX A
LABORATORY AND FIRST AID RULES

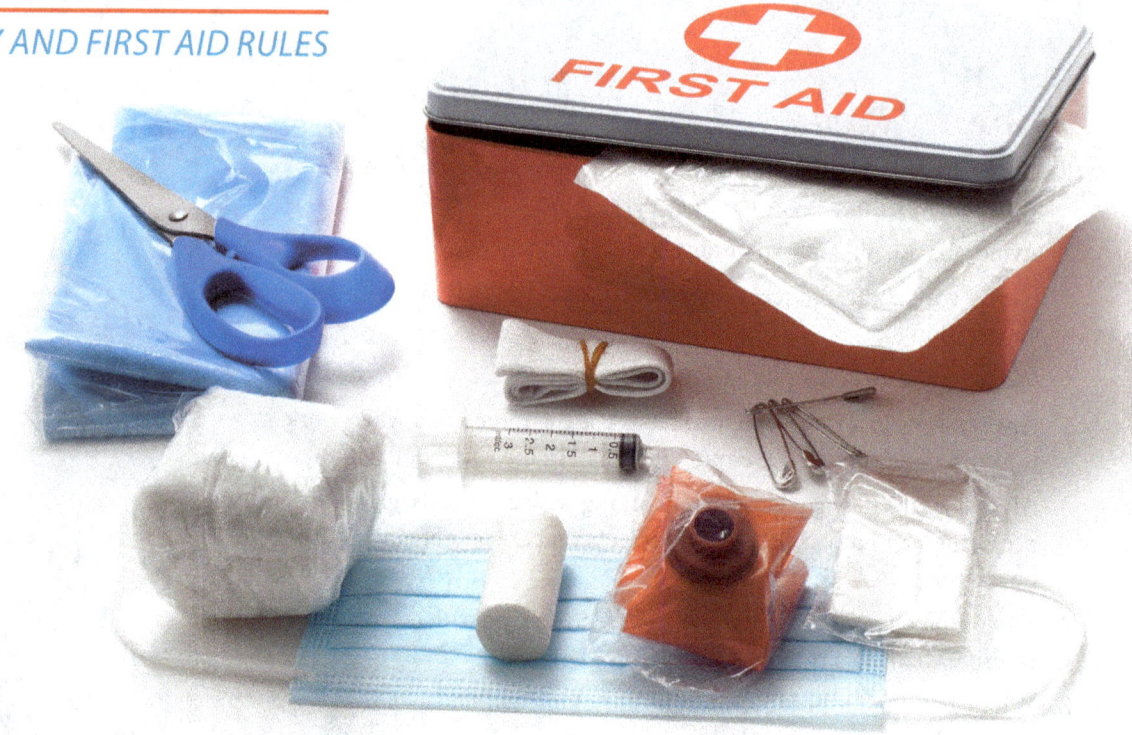

Basic First Aid

1. **Burns**
   Flush the area with cold water for several minutes. Do not apply ice.

2. **Chemical spills**
   Notify your teacher of all chemical spills.

   a. On a laboratory desk

   1. If the material is not particularly volatile, toxic, or flammable, your teacher may have you clean the spill. For liquids, use an absorbent material that will soak up the chemical. For solids, use the designated dustpan and brush. Dispose of chemicals and cleaning materials properly. Then clean the area with soap and water.

   2. If the material is volatile, toxic, or flammable, and not a large spill, ask your teacher for help. If it is a large spill, you may need to evacuate the laboratory.

   3. If a highly reactive material, such as hydrochloric acid, is spilled, your teacher will clean it up.

   b. On a person

   1. If the spill covers a large area, begin rinsing in the chemical shower, then remove all contaminated clothing and remain under the safety shower. Flood the affected body area for fifteen minutes. Obtain medical help immediately.

APPENDIX A
LABORATORY AND FIRST AID RULES

2. If the spill covers a small area, immediately flush the affected area with cold water for several minutes. Then clean the area with soap and water. Get medical attention.

3. If the chemical splashes in your eyes, immediately wash them in the nearest eyewash fountain for at least 15–20 minutes. Get medical attention.

4. In any case, if the spill is an acid, rinse the area with sodium bicarbonate or sodium carbonate solution; if it is a base, use citric or ascorbic acid solution.

3. Fire
   a. For any fire other than a contained fire, do not attempt to put it out on your own. Understand that some fires can't be put out with water.
   b. Smother a small fire in a container by covering it.
   c. If a person's clothes are on fire, remember to stop, drop, and roll—roll the person on the floor and use a fire blanket to extinguish the flames. The safety shower may also be used. *Do not use a fire extinguisher.*

4. **Swallowed chemicals**
   Determine the specific substance ingested and follow the instructions on the SDS. Contact the Poison Control Center immediately.

5. **Cuts**
   If the wound is superficial, clean it with a disinfectant, apply triple antibiotic ointment, and cover with a bandage. If it is a deep cut, seek immediate medical attention.

6. **Bites and stings**
   a. For minor bites and stings, wash the affected area with soap and water, apply hydrocortisone cream, apply a cool compress, and take an antihistamine. If there is evidence of an allergic reaction, seek immediate medical attention.
   b. If the bite is deep or if the animal is suspected of carrying rabies, seek immediate medical attention.

7. **Rashes from plants**
   a. Rashes caused by exposure to poison ivy, oak, or sumac will not appear immediately after contact; it usually takes about twenty-four hours for symptoms to show. If you suspect that you have contacted these plants, wash the area with soap and water within ten minutes of contact. Be sure to clean any clothing that has contacted plants since it may also contain the toxins.
   b. If a rash is visible, treat with calamine lotion, hydrocortisone cream, or oral antihistamines (or a combination of the three). If conditions worsen, seek medical attention.

APPENDIX B
SAFETY DATA SHEETS (SDS)

Calcium Chloride

SDS #: 196
Revision Date: July 8, 2015

Safety Data Sheet (SDS)

SECTION 1 — CHEMICAL PRODUCT AND COMPANY IDENTIFICATION

Calcium Chloride

Flinn Scientific, Inc. P.O. Box 219, Batavia, IL 60510 (800) 452-1261

Chemtrec Emergency Phone Number: (800) 424-9849

Signal Word
WARNING

Pictograms

SECTION 2 — HAZARDS IDENTIFICATION

Hazard class: Acute toxicity, oral (Category 4). Harmful if swallowed (H302). Do not eat, drink or smoke when using this product (P270).

Hazard class: Eye irritation (Category 2A). Causes serious eye irritation (H319).

SECTION 3 — COMPOSITION, INFORMATION ON INGREDIENTS

Component Name	CAS Number	Formula	Formula Weight	Concentration
Calcium chloride, anhydrous	10043-52-4	$CaCl_2$	110.98	

SECTION 4 — FIRST AID MEASURES

Call a POISON CENTER or physician if you feel unwell.

If inhaled: Remove victim to fresh air and keep at rest in a position comfortable for breathing.

If in eyes: Rinse cautiously with water for several minutes. Remove contact lenses if present and easy to do so. Continue rinsing (P305+P351+P338). **If eye irritation persists:** Get medical advice or attention (P337+P313).

If on skin: Rinse cautiously with water for several minutes (P351).

If swallowed: Rinse mouth. Contact a POISON CENTER or physician if you feel unwell.

SECTION 5 — FIRE FIGHTING MEASURES

Nonflammable, noncombustible solution.

When heated to decomposition, may emit toxic fumes.

In case of fire: Use a tri-class dry chemical fire extinguisher.

NFPA Code
None established

SECTION 6 — ACCIDENTAL RELEASE MEASURES

Sweep up the spill, place in a sealed bag or container, and dispose. Ventilate area and wash spill site after material pickup is complete. See Sections 8 and 13 for further information.

SECTION 7 — HANDLING AND STORAGE

Flinn Suggested Chemical Storage Pattern: Inorganic #2. Store with acetates, halides, sulfates, sulfites, thiosulfates and phosphates. Store in a cool, dry place. Hygroscopic. Store in Flinn Chem-Saf™ bag.

APPENDIX B
SAFETY DATA SHEETS (SDS)

SECTION 8 — EXPOSURE CONTROLS, PERSONAL PROTECTION
Wear protective gloves, protective clothing, and eye protection (P280). Wash hands thoroughly after handling (P264).

SECTION 9 — PHYSICAL AND CHEMICAL PROPERTIES

White powder, crystals, or flakes. Odorless.	Boiling point: 1670 °C
Soluble: Water and alcohol. Liberates heat in water.	Melting point: 772 °C
	Specific gravity: 2.15

SECTION 10 — STABILITY AND REACTIVITY
Avoid contact with strong acids.

Shelf life: Fair to poor, hygroscopic. See Section 7 for further information.

SECTION 11 — TOXICOLOGICAL INFORMATION

Acute effects: Irritant.	ORL-RAT LD50: 1000 mg/kg
Chronic effects: N.A.	IHL-RAT LC50: N.A.
Target organs: N.A.	SKN-RBT LD50: N.A.

SECTION 12 — ECOLOGICAL INFORMATION
Data not yet available.

SECTION 13 — DISPOSAL CONSIDERATIONS
Please review all federal, state, and local regulations that may apply before proceeding.

Flinn Suggested Disposal Method #26a is one option.

SECTION 14 — TRANSPORT INFORMATION
Shipping name: Not regulated. Hazard class: N/A. UN number: N/A.

SECTION 15 — REGULATORY INFORMATION
TSCA-listed, EINECS-listed (233-140-8).

SECTION 16 — OTHER INFORMATION
This Safety Data Sheet (SDS) is for guidance and is based upon information and tests believed to be reliable. Flinn Scientific, Inc. makes no guarantee of the accuracy or completeness of the data and shall not be liable for any damages relating thereto. The data is offered solely for your consideration, investigation, and verification. The data should not be confused with local, state, federal or insurance mandates, regulations, or requirements and CONSTITUTE NO WARRANTY. Any use of this data and information must be determined by the science instructor to be in accordance with applicable local, state, or federal laws and regulations. The conditions or methods of handling, storage, use, and disposal of the product(s) described are beyond the control of Flinn Scientific, Inc. and may be beyond our knowledge. FOR THIS AND OTHER REASONS, WE DO NOT ASSUME RESPONSIBILITY AND EXPRESSLY DISCLAIM LIABILITY FOR LOSS, DAMAGE OR EXPENSE ARISING OUT OF OR IN ANY WAY CONNECTED WITH THE HANDLING, STORAGE, USE OR DISPOSAL OF THIS PRODUCT(S). N.A. = Not available, not all health aspects of this substance have been fully investigated.

N/A = Not applicable

Consult your copy of the Flinn Science Catalog/Reference Manual for additional information about laboratory chemicals.

Revision Date: July 8, 2015

APPENDIX B
SAFETY DATA SHEETS (SDS)

Hydrochloric Acid

SDS #: 395.2
Revision Date: December 3, 2018

Safety Data Sheet (SDS)

SECTION 1 — CHEMICAL PRODUCT AND COMPANY IDENTIFICATION

Hydrochloric Acid Solution 0.1 M - 2.4 M

Flinn Scientific, Inc. P.O. Box 219, Batavia, IL 60510 (800) 452-1261

Chemtrec Emergency Phone Number: (800) 424-9849

Signal Word
DANGER

Pictograms

SECTION 2 — HAZARDS IDENTIFICATION

Hazard class: Skin corrosion or irritation (Category 1). Causes severe skin burns and eye damage (H314). Do not breathe mist, vapors, or spray (P260).

Hazard class: Specific target organ toxicity, single exposure; respiratory tract irritation (Category 3). May cause respiratory irritation (H335).

Hazard class: Acute toxicity, oral (Category 5). May be harmful if swallowed (H303).

Industrial exposure to hydrochloric acid vapors and mists is listed as a known human carcinogen by IARC (IARC-1).

SECTION 3 — COMPOSITION, INFORMATION ON INGREDIENTS

Component Name	CAS Number	Formula	Formula Weight	Concentration
Hydrochloric acid	7647-01-0	HCl	36.46	2-10%
Water	7732-18-5	H_2O	18.00	90-98%

SECTION 4 — FIRST AID MEASURES

Call a POISON CENTER or physician if you feel unwell (P312).

If inhaled: Remove victim to fresh air and keep at rest in a position comfortable for breathing (P304+P340). **If in eyes:** Rinse cautiously with water for several minutes. Remove contact lenses if present and easy to do so. Continue rinsing (P305+P351+P338). Immediately call a POISON CENTER or physician (P310). **If on skin or hair:** Immediately remove all contaminated clothing. Rinse skin with water (P303+P361+P353). Wash contaminated clothing before reuse (P363). **If swallowed:** Rinse mouth. Do NOT induce vomiting (P301+P330+P331).

SECTION 5 — FIRE FIGHTING MEASURES

Nonflammable, noncombustible solution.

In case of fire: Use a tri-class dry chemical fire extinguisher.

NFPA Code
None established

SECTION 6 — ACCIDENTAL RELEASE MEASURES

Ventilate area and contain the spill with sand or other inert absorbent material, neutralize with sodium bicarbonate or calcium hydroxide, and deposit in a sealed bag or container. See Sections 8 and 13 for further information.

SECTION 7 — HANDLING AND STORAGE

Flinn Suggested Chemical Storage Pattern: Inorganic #9. Store with acids, except nitric acid.

Store in a dedicated acid cabinet; if an acid cabinet is not available, store in Flinn Saf-Cube™. Keep container tightly closed (P233). Use only in a hood or well-ventilated area (P271).

APPENDIX B
SAFETY DATA SHEETS (SDS)

SECTION 8 — EXPOSURE CONTROLS, PERSONAL PROTECTION

Wear protective gloves, protective clothing, and eye protection (P280). Wash hands thoroughly after handling (P264). Use only in a hood or well-ventilated area (P271).

Exposure guidelines: (as concentrated HCl) Ceiling 5 ppm (OSHA); Ceiling 2 ppm (ACGIH); IDLH 50 ppm. Irritation threshold is ~ 5 ppm so any irritation is sign of exposure (per OSHA)

SECTION 9 — PHYSICAL AND CHEMICAL PROPERTIES

Clear, colorless liquid. Acrid, chlorine odor.	pH: < 1
Soluble: Water and alcohol	

SECTION 10 — STABILITY AND REACTIVITY

Avoid contact with strong oxidizers, bases, amines, and alkali metals. Corrodes metal including stainless steel.

Shelf life: Good, if stored properly.

SECTION 11 — TOXICOLOGICAL INFORMATION

Acute effects: Eye and skin corrosion. Respiratory irritation, coughing.	ORL-RBT LD50: 900 mg/kg (as concentrated HCl)
Chronic effects: Corrosive to teeth.	IHL-RAT LC50: 3124 ppm/1 hour (as concentrated HCl)
Target organs: Respiratory tract, teeth, skin, eyes.	SKN-RBT LD50: N.A.

SECTION 12 — ECOLOGICAL INFORMATION

Does not biodegrade in soil, may be toxic to aquatic life.

SECTION 13 — DISPOSAL CONSIDERATIONS

Please review all federal, state and local regulations that may apply before proceeding.

Flinn Suggested Disposal Method #24b is one option.

SECTION 14 — TRANSPORT INFORMATION

Shipping name: Hydrochloric acid. Hazard class: 8, Corrosive. UN number: UN1789.

SECTION 15 — REGULATORY INFORMATION

TSCA-listed, EINECS-listed (231-595-7), RCRA code D002.

SECTION 16 — OTHER INFORMATION

This Safety Data Sheet (SDS) is for guidance and is based upon information and tests believed to be reliable. Flinn Scientific, Inc. makes no guarantee of the accuracy or completeness of the data and shall not be liable for any damages relating thereto. The data is offered solely for your consideration, investigation, and verification. The data should not be confused with local, state, federal or insurance mandates, regulations, or requirements and CONSTITUTE NO WARRANTY. Any use of this data and information must be determined by the science instructor to be in accordance with applicable local, state, or federal laws and regulations. The conditions or methods of handling, storage, use and disposal of the product(s) described are beyond the control of Flinn Scientific, Inc. and may be beyond our knowledge. FOR THIS AND OTHER REASONS, WE DO NOT ASSUME RESPONSIBILITY AND EXPRESSLY DISCLAIM LIABILITY FOR LOSS, DAMAGE OR EXPENSE ARISING OUT OF OR IN ANY WAY CONNECTED WITH THE HANDLING, STORAGE, USE OR DISPOSAL OF THIS PRODUCT(S).

N.A. = Not available, not all health aspects of this substance have been fully investigated.

N/A = Not applicable

Consult your copy of the Flinn Science Catalog/Reference Manual for additional information about laboratory chemicals.

Revision Date: December 3, 2018

APPENDIX C
LABORATORY EQUIPMENT

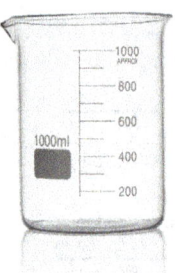
beaker

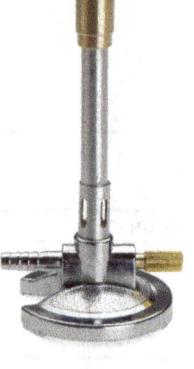
Bunsen burner

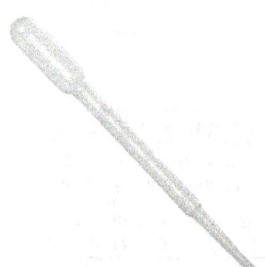
disposable pipette

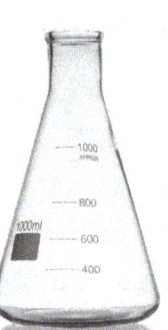
Erlenmeyer flask

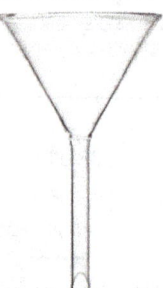
filter funnel

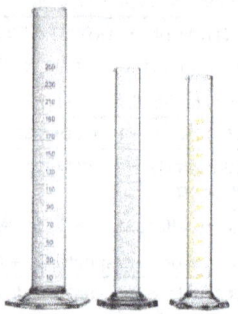
graduated cylinder

hot plate

iron ring

laboratory balance (electronic)

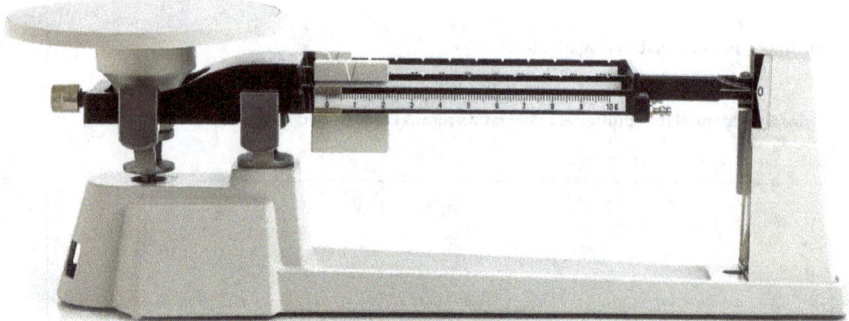
laboratory balance (triple beam)

mortar and pestle

APPENDIX C
LABORATORY EQUIPMENT

pipette

ring stand

safety goggles

spatula

stirring rod

test tube

test tube holder

test tube rack

wash bottle

watch glass

wire gauze (on a stand)

Appendix C 245

APPENDIX D
LABORATORY TECHNIQUES

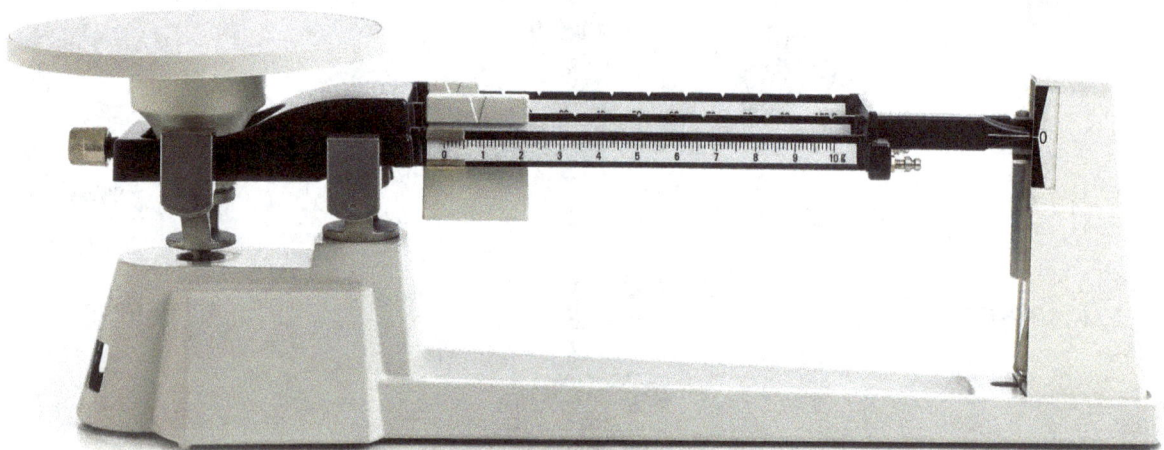

Triple-beam balance

Measuring Mass

Using Mechanical Balances

The mass of a substance can be determined in the laboratory with the use of a mechanical balance. Several kinds of mechanical balances are common, but all of them operate on the same principles. To use a mechanical balance properly, follow the steps given below.

1. Place the balance on a smooth, level surface.
2. Keep the balance pan(s) clean and dry. Never put chemicals directly on the metal surface of the pan(s). When massing chemicals, place substances on a sheet of weighing paper, watch glass, or in a beaker.
3. Check the rest point of the empty balance. To do this, remove all masses from the pans and slide all movable masses to their zero positions. If the balance beam swings back and forth, note the central point of the swing. You do not have to wait until the beam stops swinging completely. If the central point lies more than two divisions from the marked zero point, ask your teacher to adjust the balance. *Do not adjust the balance yourself!*

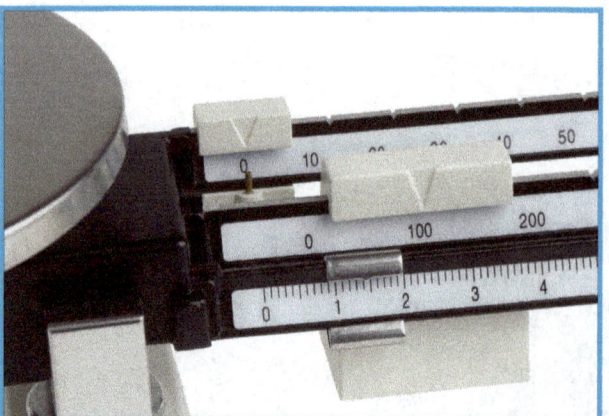

4. Place the substance to be massed on the pan and adjust the sliding masses. Move the largest masses first, and then make final adjustments with the smaller masses. If the beam has *detents* for the sliding masses, make sure that the masses rest in the detent. The sum of all the readings is the mass of the object. The mass of the sample in the image shown at left should be read as 101.43 g.

APPENDIX D
LABORATORY TECHNIQUES

Using an Electronic Balance

Electronic balances are generally faster and easier to use than their mechanical counterparts. To use an electronic balance properly, follow the instructions given below.

1. Place the balance on a smooth, level surface.
2. Keep the balance pan(s) clean and dry. Never put chemicals directly on the metal surface of the pan(s). When massing chemicals, place substances on a sheet of weighing paper, on a watch glass, or in a beaker.
3. Turn the balance on. Place the container or weigh paper that will hold the substance to be massed and make sure that there is a reading of 0 by pushing the **Tare** button.
4. Place your substance on the paper or in the container. Add the desired substance until you have reached the appropriate mass.

Electronic balance

Using a Laboratory Burner

Laboratory burners are a common source of heat in laboratories. They are popular because they give a hot flame and burn clean, readily available natural gas or butane. Laboratory burners work well because they mix gas with the correct amount of air to produce the most heat. If air is not mixed with the gas before it burns, not all the gas will burn and the flame will not be as hot. If too much air is mixed with the gas, the air will extinguish the flame.

If the burner doesn't have its own gas supply, connect it to the desk gas line with a rubber hose. Open the main gas valve and light the burner with a match or flint. Adjustments to the flame should be made from the needle and air valves. If the burner lights but the flame immediately goes out, try increasing the gas flow at the needle valve. A yellow flame signifies that insufficient air is mixing with the gas. A flame that makes a noise like a roaring wind means that too much air is entering the barrel. This extra air may cool the flame or extinguish it entirely. To get the best flame possible, rotate the barrel until the flame is entirely blue and two distinct zones appear. Place objects to be heated at the tip of the inner blue zone for quick heating. Sometimes the flame strikes back; that is, it enters the barrel and comes out the bottom. If this happens, do not panic. Turn off the gas supply and readjust the burner so that less air enters the barrel.

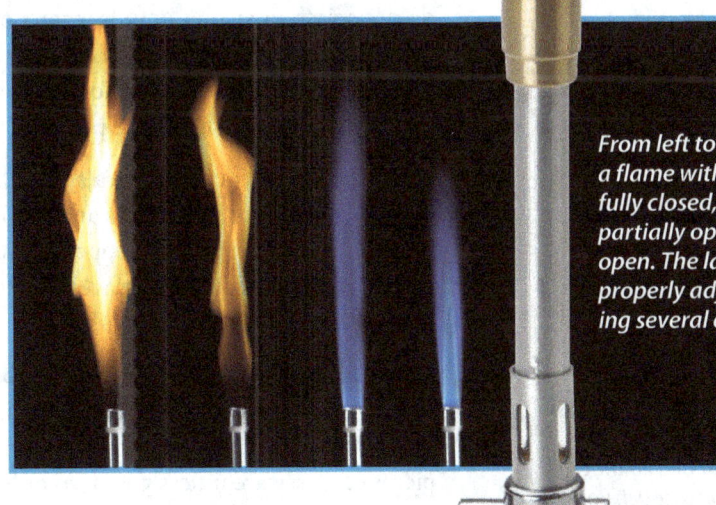

From left to right, notice a flame with the air valve fully closed, mostly closed, partially open, and fully open. The last flame is properly adjusted, showing several cones of flame.

Laboratory burner

APPENDIX D

LABORATORY TECHNIQUES

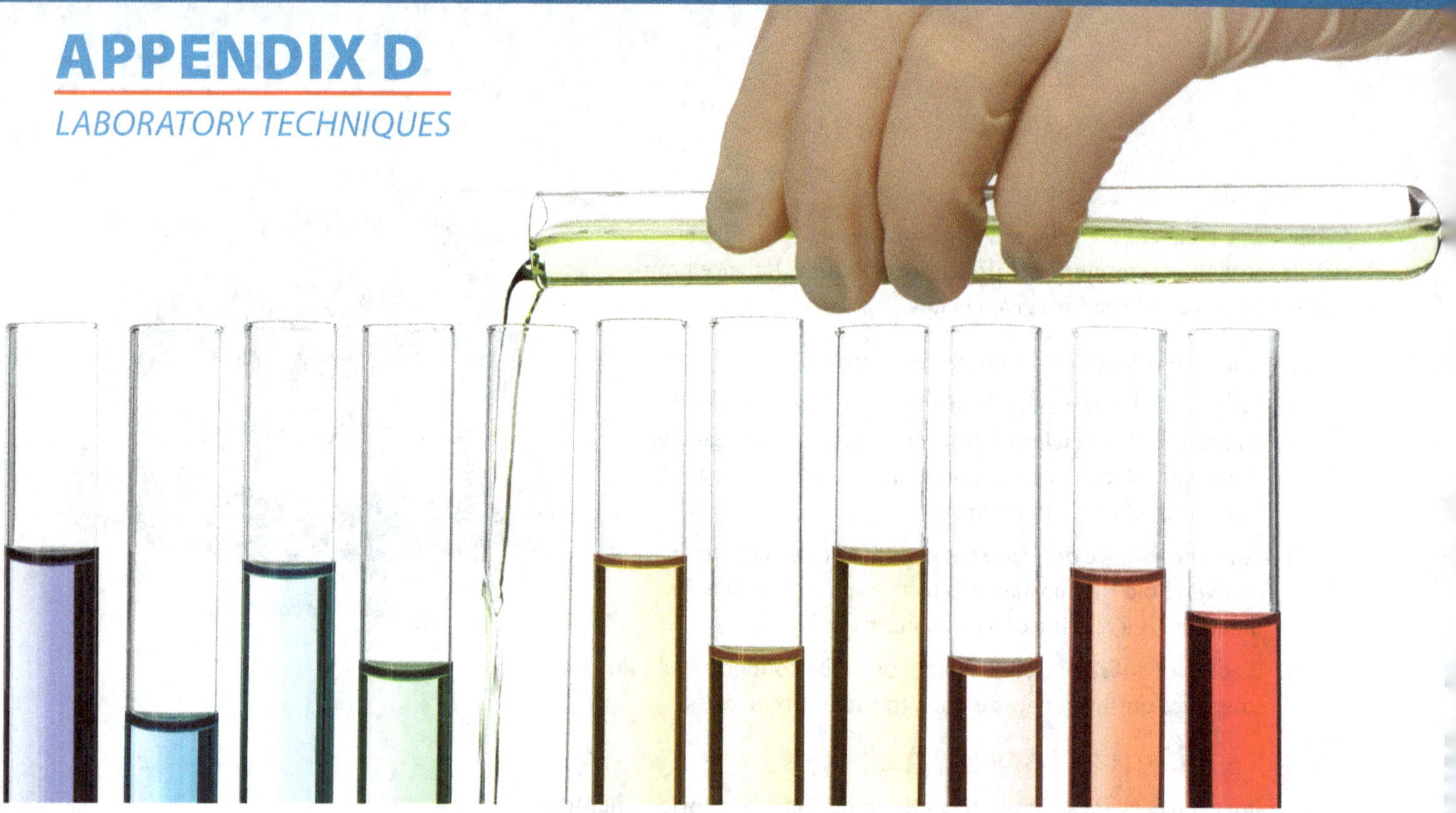

Measuring Chemicals

Proper technique for handling liquids is essential if you are to remain safe, keep reagents pure, and obtain accurate measurements. For increased safety, do not splash or spatter liquids when pouring. Pour them slowly down the insides of test tubes, graduated cylinders, and beakers. If anything is spilled, wipe it up promptly. To keep liquids from running down the outside of the container from which you are pouring, pour the liquid down a stirring rod.

In order to keep liquid chemicals pure, keep stirring rods out of the stock supply. Do not let the stoppers and lids become contaminated while you are pouring. Instead, hold the stopper between your fingers. If you must put a lid down, keep the inside surface from touching the surface of the laboratory table, as shown in the image at left.

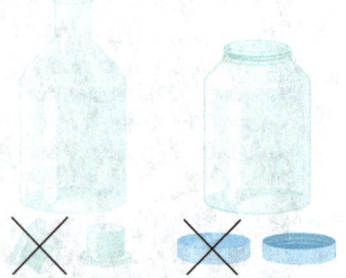

Don't let the part of a lid that touches a chemical touch your laboratory bench.

Accurate measurements of liquids can be made in burets, graduated cylinders, and volumetric flasks. You should measure volumes in these pieces of glassware unless you need only a rough approximation. If the liquid surface curves upward (i.e., the liquid is pulled up the sides of the cylinder), read the measurement at the bottom of the curved surface—the *meniscus* (right). In some cases the meniscus will curve downward, in which case you should read the volume at its highest point.

© 2020 BJU Press. Reproduction prohibited.

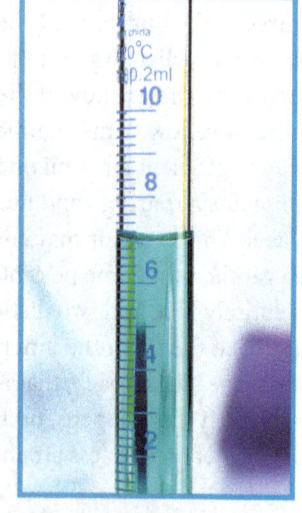

APPENDIX D
LABORATORY TECHNIQUES

To measure out solids, (1) scoop out a little of the sample onto a sheet of weighing paper with a spatula; (2) gently tap the spatula until the desired amount falls off, and (3) cup the paper to pour the sample into a test tube.

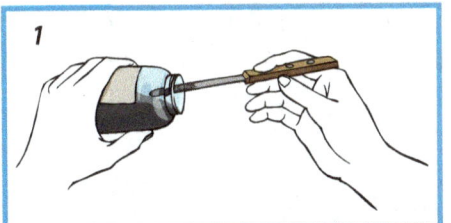

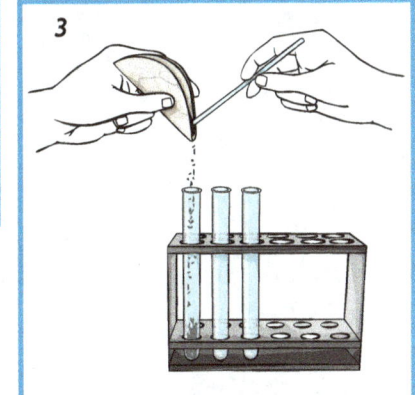

Using a Thermometer

When using a thermometer, make sure that you are using one that has the proper temperature range for the experiment that you will be doing. Always hold a thermometer when it is in a container. *Never leave a thermometer standing upright in a container; the container is likely to get knocked over.*

When taking a reading, position the thermometer bulb just above the bottom of the container. If the bulb touches the container, your readings will be inaccurate.

If a thermometer breaks, alert your teacher and do not touch the inner contents. Some thermometers contain mercury. The spilled mercury may look fascinating, but it is toxic and can be absorbed through the skin.

Separating Liquids and Solids

Several experiments require that you separate a solid from a liquid. The most common method of separation—filtering—involves passing the solution through a fine sieve, such as filter paper. The paper allows the liquid and dissolved particles to pass through but catches undissolved particles.

The filter paper must be folded to fit the funnel. Fold it in half and then tear off the corner as shown (red circle) at right. Fold it again at an angle slightly less than 90° to the first fold. Open the paper to form a cone—half of the cone should have three layers of paper, and the other half should have one. Place the cone in a funnel and wet the paper with a few drops of distilled water to hold it in place. Seal the edge of the paper against the edge of the funnel so that none of the solution can go down the spout without going through the paper.

Decanting is a quick method that is often acceptable for separating a liquid from a solid. To decant, allow the solid to settle and then gently pour the liquid off the top of the residue (right). Avoid causing turbulence that could mix the solid with the liquid. Sometimes the solid residue left in the container is rinsed with distilled water and decanted a second time to make sure that all the liquid is separated from the precipitate.

APPENDIX E
STEM DESIGN PROCESS

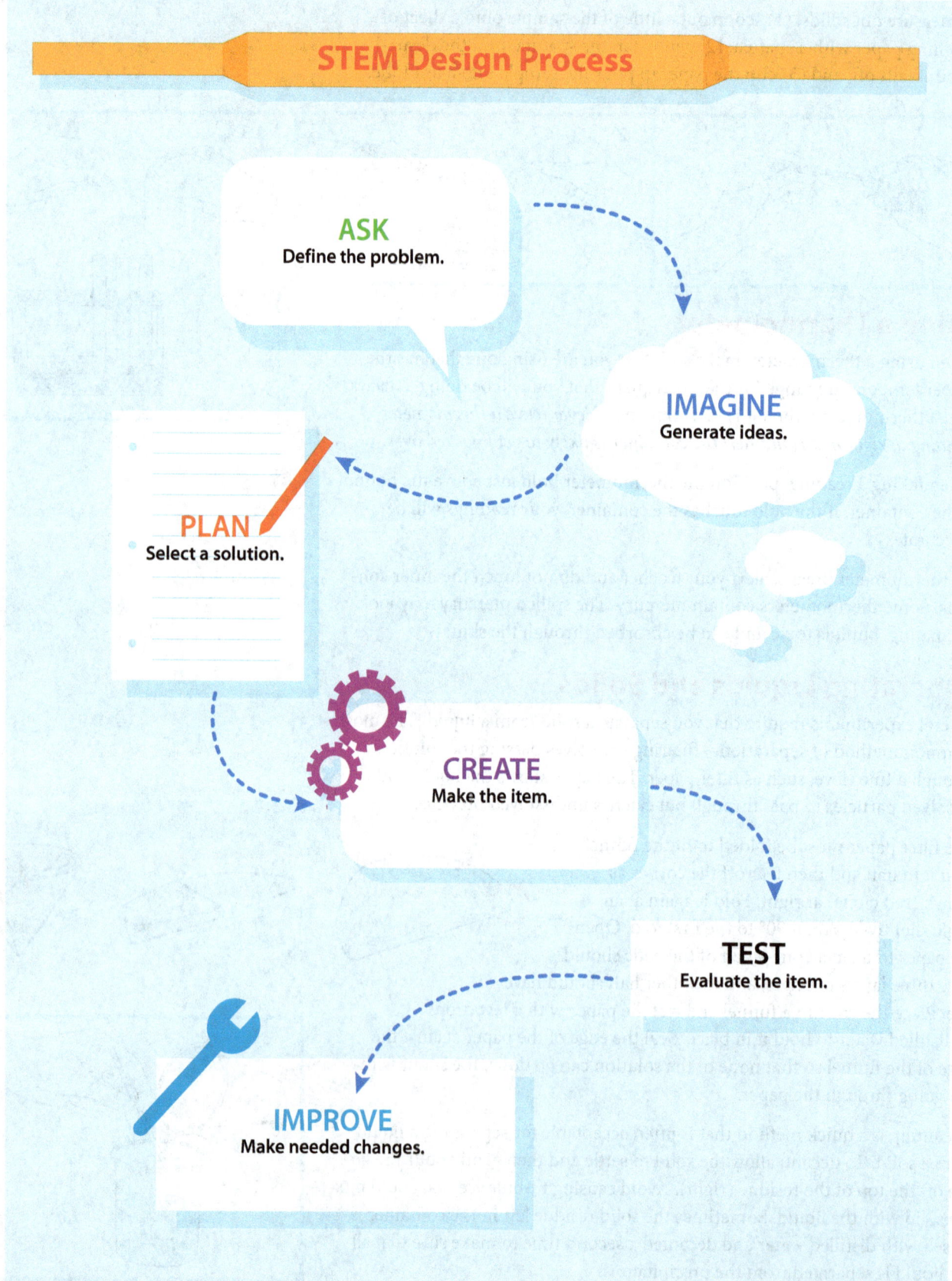

APPENDIX F
INVESTIGATING WITH LABDISC

Getting Started

Making models is one of the most important things that scientists do. Models can be physical objects, but most are either mathematical equations or visual representations of data. For example, if a scientist wanted to model human growth from birth to adulthood, he would probably graph height versus age, as in the graph at right. Visual models are especially useful because we are usually much better at interpreting pictures than tables of numbers.

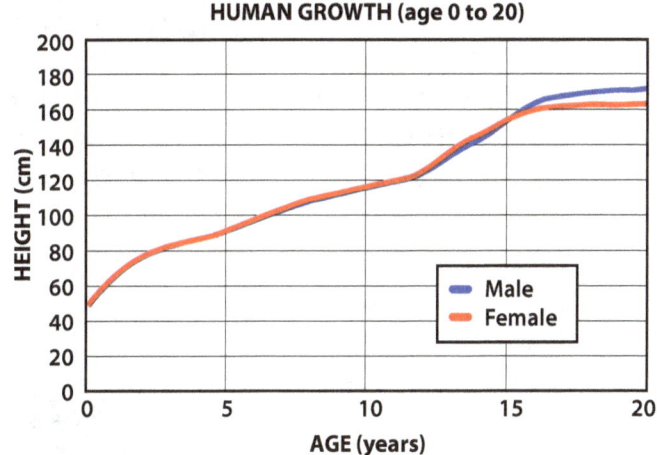

To make models, scientists need *data*, which is usually numerical. Most data comes from *instruments*, devices that measure physical phenomena. Collecting data by hand can be a very time-consuming process, which is one reason that many high-school lab activities don't require students to collect large quantities of data. Additionally, to make an accurate model there needs to be a great quantity of data, which requires time. Imagine collecting 500 data points and then having to graph that data by hand! While graphing software could make the task easier, you still would have to type in the raw data by hand. What's the solution?

Probeware to the Rescue!

In the early 1990s, several companies began to sell educational-grade lab sensors that connected to a personal computer, turning it into a powerful and flexible instrument. Using these systems, students could gather large data sets in many different areas of science, including earth science, biology, chemistry, and physics. Modeling was easy too since the data was already in the computer.

By the mid-2000s, the technology had matured and became affordable and portable. Instead of a personal computer, the sensors now connect to small, handheld devices that include built-in modeling software. This portability makes it possible to go into the field and collect data. We call this kind of technology *probeware*.

Today there are even probeware systems that have the sensors built into the handheld device. One such company is Globisens, which makes the Labdisc. The Labdisc data can be analyzed with software that is available online. To download the program, visit TeacherToolsOnline.com. Under Additional Resources, click on *Globisens*.

APPENDIX F
INVESTIGATING WITH LABDISC

Labdisc Hardware Overview
LABDISC GENSCI—CONTENTS

1. Labdisc data logger
2. Labdisc AC charger
3. USB cable
4. software flyer
5. banana cables
6. Quick Start guide
7. temperature probe
8. air pressure tube

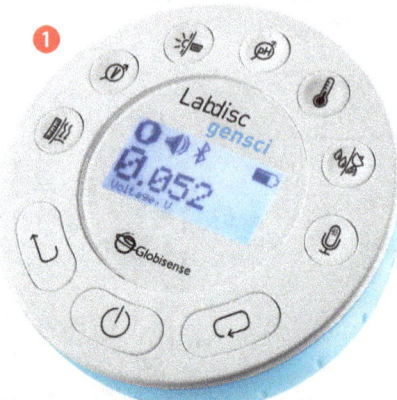

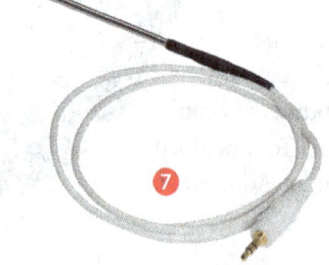
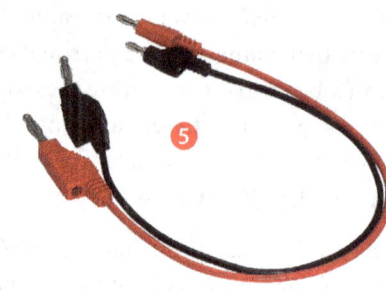
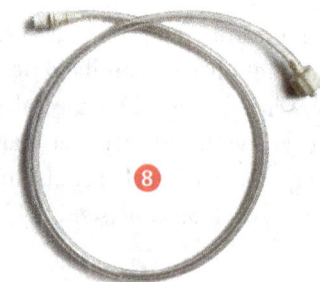

APPENDIX F
INVESTIGATING WITH LABDISC

PORTS AND CONTROLS

1. On/Off and Escape key
2. Scroll key
3. Select key
4. Sensor selection keys
5. graphical display
6. rotating ring
7. USB port
8. plastic leg
9. M5 screw insert
10. distance sensor
11. microphone, sound level sensors
12. relative humidity sensor
13. external temperature input
14. pH input
15. light sensor, universal input
16. current and voltage sensor
17. air pressure sensor
18. GPS sensor (internal)

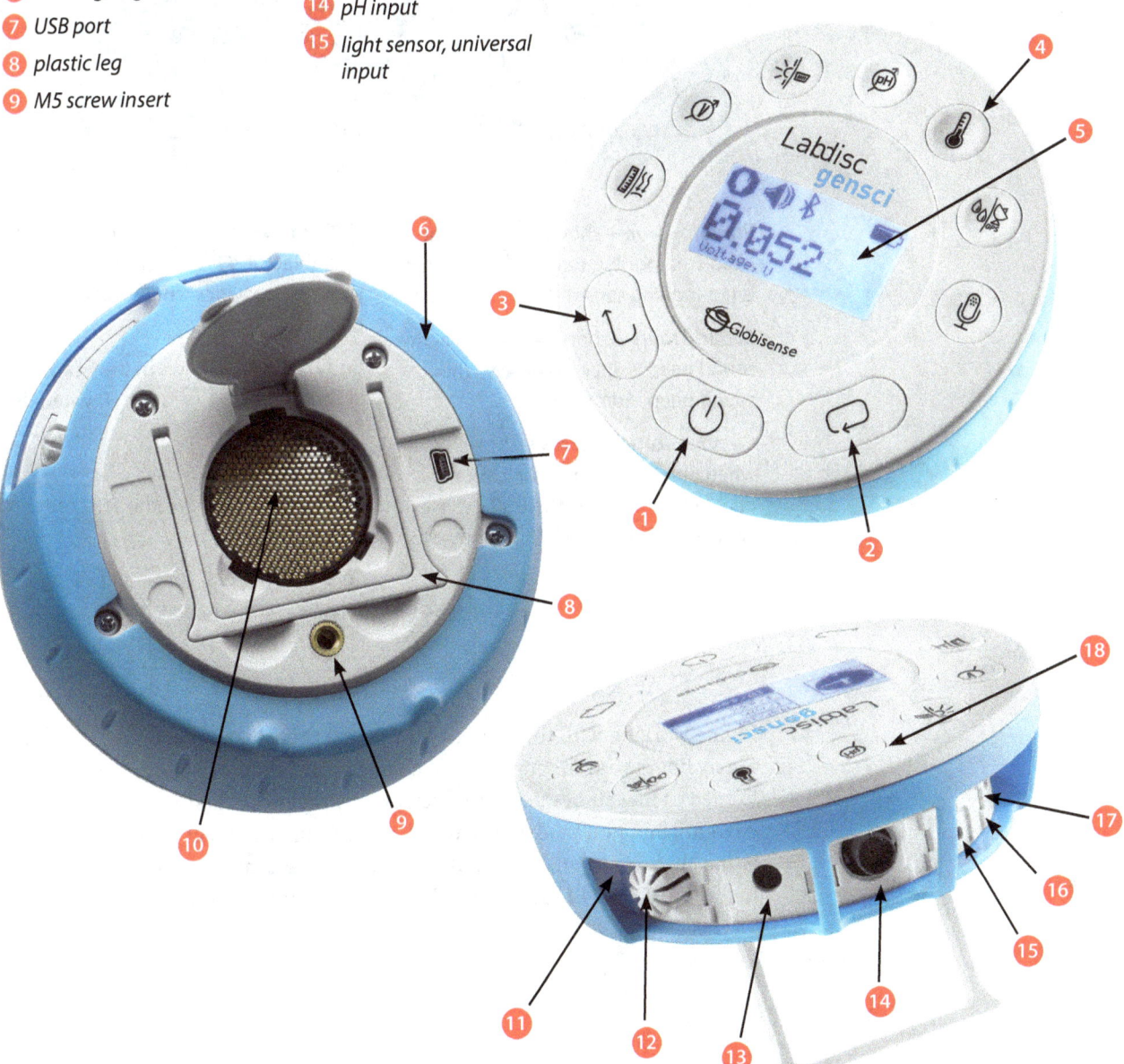

Appendix F 253

APPENDIX F
INVESTIGATING WITH LABDISC

Labdisc Display

The Labdisc LCD display allows users to see the different sensor readings and to set up or reconfigure the Labdisc parameters.

Run/Stop Icon—shows 🏃 when the Labdisc is logging data or ⏺ when it is not logging sensor data

Sound Status—shows 🔊 when active sound beep is enabled or 🔇 when sound beep is disabled

Communication Status—shows ✱ when Bluetooth® communication is enabled or ⚡ when the USB cable is connected from the host computer to the Labdisc

GPS Status—shows 📡 when the GPS is enabled or 📶 when locked to GPS satellites; provides valid positioning parameters

Battery Level—shows battery capacity at three levels 🔋🔋🔋 or ⚡ when the Labdisc is connected to the external charger

Sensor Value—shows the selected sensor value

Sensor Name and Unit—shows the selected sensor name and unit

Charging the Labdisc

Before working with the Labdisc for the first time, the unit should be charged for six hours with the supplied 6V charger. The Labdisc charging input is located to the left of the On/Off key. Simply move the rotating ring until the charging input on the Labdisc is exposed and then connect the charger plug to the charging input. The Labdisc charger will accept any input voltage ranging from 100 to 240 VAC 50/60 Hz, making it functional worldwide.

APPENDIX F
INVESTIGATING WITH LABDISC

Pairing the Labdisc via Bluetooth

Before you can connect the Labdisc via Bluetooth, you will need to pair the devices.

A Turn on the Labdisc by pressing the **On/Off** key.

B *Enable Bluetooth*—Press **Scroll** key twice to select the configuration menu. Press the **Select** key to open the Configuration Menu. Press the **Scroll** key to select the Bluetooth configuration menu and then press the **Select** key to open the Bluetooth Menu. Press the **Scroll** key to select **BT enable** and then press the **Select** key to enable Bluetooth. Press the **Escape** key three times to return to the main screen.

C If you have not already downloaded the GlobiLab data analysis program, visit TeacherToolsOnline.com. Under Additional Resources, click on *Globisens*.

D Open the GlobiLab software and right-click on the **Bluetooth** icon in the bottom right corner of the screen. Select **Find More Labdiscs and Sensors**. The computer will search and find any Bluetooth devices in the area. In the pop-up window, select your Labdisc. If there are multiple Labdiscs running in the area, be sure that the number on the back of your Labdisc matches the Labdisc that you select from this list.

E Press the **Scroll** key until you hear a long beep sound and the Labdisc displays "BT PAIRING."

F On your computer, select **Enter the device pairing code** and click **Next**.

G In the next dialog box, enter "1234" as the pairing code and click **Next**.

H Wait for the computer to finish the process and announce, "Your device is ready to use."

APPENDIX F
INVESTIGATING WITH LABDISC

Setting Up the Labdisc
FROM THE LABDISC

I Turn on the Labdisc by pressing the **On/Off** key.

J Move the rotating ring to expose the sensors. Connect any external sensors that you might be using.

K *Select Sensors*—Press the **Scroll** key to open the sensor menu. Press the **Select** key to open the Setup Menu, and then choose the **Set Sensor** icon using the **Select** key again. Select the sensors that you want to use by pushing the buttons on the perimeter of the Labdisc. Some buttons control more than one sensor. Push the button multiple times to scroll through the available options. When the sensor(s) that you want is (are) highlighted, push the **Escape** key to return to the Setup Menu. *Note*: Some buttons can activate only one sensor at a time because of the external connections needed by the sensors. When using any of the Labdisc sensors, be sure to move the rotating ring so that the sensor is exposed to the environment during the experiment unless otherwise noted.

L *Set Sampling Rate*—Press the **Escape** key to return to the previous menu and use the **Scroll** key to move to the **Sampling Rate** icon and push the **Select** key. Use the **Scroll** key to move through the options. Use the **Select** key to choose the rate at which you want to select samples. Press the **Select** key to confirm your choice. Options include manual, 1/second, 10/second, 100/second, 1000/second, and 25 000/second. The rate of collection by some sensors is limited, so you may not see all the options for every sensor. *Note*: If you choose to sample manually, skip Step M.

M *Set the Number of Samples*—Press the **Escape** key to leave the previous menu and use the **Scroll** key to move to the **Number of Samples** icon and push the **Select** key. Use the **Scroll** key to choose the number of samples you want to collect. Options include 10, 100, 1000, and 10 000. Press the **Select** key to confirm your choice. Press the **Escape** key three times to exit the Setup Menu.

APPENDIX F
INVESTIGATING WITH LABDISC

THROUGH THE GLOBILAB DATA ANALYSIS PROGRAM

N Open the GlobiLab software and turn on the Labdisc by pressing the **On/Off** key .

O Connect the Labdisc to the computer with the USB cord or via Bluetooth. The first time you connect via Bluetooth, you will pair the devices (see above).

Bluetooth: Right-click the **Bluetooth** icon in the lower right corner of the program. Find your paired Labdisc and click on it. The computer will connect to the Labdisc and turn the **Bluetooth** icon blue.

USB: Connect the computer and Labdics with the USB cable. The **USB** icon will turn blue.

P Click on to configure the Labdisc.

Q Select sensors that you want to use in the Logger Setup window. Different sensors will appear for different models of Labdisc, depending on what sensors are available. If you are using a Globisens brand external sensor attached to the Universal Adapter port, it will show up in the Setup Menu after you have attached it. If you are using a sensor from another company, you will need to select it from the drop-down menu.

R Click the drop-down arrow under Rate to select the sample frequency.

S Click the drop-down arrow under Samples to select the number of samples that you want to collect.

T Click **Exit** to complete the setup.

U *Begin Collecting Data*—When you are ready, press the **Select** key to begin the data collection.

APPENDIX F
INVESTIGATING WITH LABDISC

Data Analysis Software
DOWNLOAD YOUR DATA

V Start the Globilab software on your computer and connect the Labdisc to the computer with the USB cord or via Bluetooth. The first time you connect via Bluetooth, you will pair the devices (see above).

Bluetooth: Right-click the **Bluetooth** icon in the lower right corner of the program. Find your paired Labdisc and click on it. The computer will connect to the Labdisc and turn the **Bluetooth** icon blue.

USB: Connect the computer and Labdisc with the USB cable. The **USB** icon will turn blue.

W Click on the **Download** button to download the last data collection into the Globilab software. If you have collected more than one data set, click the drop-down arrow and select the **Download menu** option. This will allow you to select from a list of prior data collections. Save the data collection by clicking the **Save** icon. Be sure to clearly label your data collection so that you don't confuse it with the other data sets.

CLEAR THE LABDISC MEMORY

X Right-click on the numbers 0/127 to the far right and select **Delete last** from the pop-up menu to remove the most recent data collection from the Labdisc, or you can select **Delete all** to remove all data from the Labdisc's memory. Be sure to save any data that you want to use later before clearing memory. You cannot clear the Labdisc memory directly from the Labdisc.

Using the Globilab Software (PC/Mac)
THE MENU/TOOL BAR

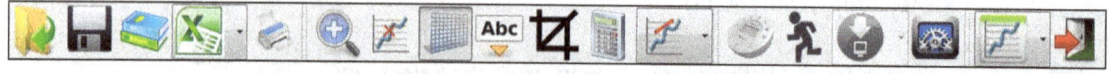

—opens graphs that show experiment data previously collected and saved

—saves data from a current or downloaded experiment

—opens PDF files of lesson plans

—includes two options: export the data in chart form into an Excel.csv file or save a screen shot of the experiment graph as a PDF file

APPENDIX F
INVESTIGATING WITH LABDISC

—prints a copy of the experiment graph

—allows you to enlarge a section of the graph

Use a left click and drag the mouse to make a box over the area that you want to enlarge. You can do this multiple times, zooming in closer and closer on a chosen section. To return to the original size, right-click on the graph. The Zoom toggles on and off, so be sure to turn it off before manipulating the graph when it's zoomed in or before selecting other buttons, like the Marker or Annotate buttons.

—allows you to place markers along a graph line using a left mouse click

A pop-up window gives you detailed information about the data at that specific point in time. Hold down the left mouse button to slide the marker along the graph line to see details at different points in time, or you can drag them to a different line to see information about other types of data in the same collection. Only two markers can be set on the screen at one time. By double-clicking on the marker, you can lock it in place. Markers are also used to set the parameters of an area that you want to crop from within the graph. To remove a marker, right-click on it and select **Yes** when asked if you want to delete the marker. The marker button toggles on and off. To exit marker mode, click the marker button again.

—turns the background grid on and off

—allows you to add labels to your graph to explain what is happening on the graph.

There are two types of annotations: a rectangle and a bubble shape that can be linked to specific places on the graph. You can also add pictures to your annotations by clicking the **add image** button. You can add an unlimited number of annotations to a graph. To remove an annotation, right-click on it and select **Yes** when asked if you want to delete the annotation. The annotation button toggles on and off.

—allows you to select a specific portion of the graph to be extracted and saved separately

To crop a section of the graph, use the marker button to set a marker at the beginning and the end of the area that you want to save. Then select the crop button. All other data will be removed *permanently* from the graph.

—calculates statistics for the graphed data, including the maximum and minimum values for a specific data collection as well as the average reading, the standard deviation, and the number of samples and sampling rate.

If you have collected multiple types of data in a single experiment, you'll need to click on the data labels in the upper right-hand corner of the graph to select which type of data you want to see in the statistics.

APPENDIX F
INVESTIGATING WITH LABDISC

Mathematical Functions. Globilab allows you to apply many different mathematical functions to your data. This drop-down menu will allow you to select from the icons below.

—allows you to find the slope at any point on the graph

—displays the best linear line to fit the graph between the two markers

A pop-up box will display the mathematical formula for the line between the two markers in the $y = ax + b$ format.

—provides the best parabolic line to fit the graph between the two markers

The pop-up box will display the mathematical formula for the line between the two markers in the $y = ax^2 + bx + c$ format.

—splits the display to show the original measurements in a time scale in the top window and to show its harmonics on a frequency scale in the lower window

—averages a graph

Every sample is an average of the two before readings and the two after readings and smooths the graph to reduce the variability of a "noisy" graph.

—measures the sensitivity to change of a quantity in one data source that is determined by another quantity or another data source

—allows you to select sensors and set up the desired collection rate and collection size from the computer rather than from the Labdisc itself

In order to use the **Setup** button, the Labdisc must be connected to the computer or tablet by either a Bluetooth or USB connection.

—starts and stops the data collection

This button will work only if the Labdisc is connected to the computer or tablet by a Bluetooth or USB connection.

—offers two options: download the data from the last experiment only or select and download data from any experiment remaining on the Labdisc

—provides options to set up the physical running of the Labdisc, including sound on/off, temperature in Celsius/Fahrenheit, Bluetooth on/off, GPS enabled/disabled, and the language used by the Labdisc

Display Options Menu. This allows you to select the format for data display. It includes the following.

—**Line Graph:** if there is data of more than one type displayed, you can set the scale on the left for different active sensors by left-clicking on the sensor name in the legend in the upper right-hand corner of the

APPENDIX F
INVESTIGATING WITH LABDISC

screen. A single right click on the sensor name will change the line to individual data points. A second right click will hide the sensor data from the graph. A third right click will return the sensor graph to a line. To change the color of a line, left-click on the line and select a new color.

—**Bar Graph:** shows data from only one sensor at a time

To change which data is displayed, left-click the sensor name in the upper left-hand corner of the screen.

—**Data Table:** shows data in a chart format

This can be used to view exact measurements for every collection. If you have an outlier data point, it can be eliminated or corrected in this format.

—**Line Graph and Data Table:** shows both a line graph and a data table simultaneously

—**Meters:** best used when the Labdisc is connected to the computer during data collection since they show changes in the data values as they are happening

There are several different options for meters including Full Dial, Half Dial, Vertical Bar, Horizontal Bar, and Numeric Readout Red, Blue & Green Colorimeter Bulb. To change between different meter types or to adjust the range of a meter, left-click on the meter to open a pop-up window where you can select options. You can select to view 1, 2, 4, or 6 meters by clicking on the **Meter** icons at the bottom of the screen.

—**Google Earth Map:** allows you to overlay a Google Earth map of the area where your experiment was conducted with the data from the sensors in that experiment

You must have an internet connection for this to work properly.

CUSTOMIZING THE DISPLAY WINDOW

Setting the graph title. To customize the graph with a title, double-click on the graph title to get a pop-up window. You can then add the title to your graph. The default title will be the date and time that the experiment was conducted.

Setting y-axis legend. To set the y-axis for a specific sensor, left-click on the sensor name in the upper left-hand side of the screen (A) to activate that sensor's reading.

Adjusting the Scale. Right-click on the y-axis scale to get a pop-up window. Type in a minimum number slightly lower than your lowest reading and a maximum number slightly higher that your highest reading. This will give the greatest detail on your graph.

PHOTO CREDITS

Key: (t) top; (c) center; (b) bottom; (l) left; (r) right; (bg) background; (i) inset

Cover
Picsfive/Shutterstock.com

Front Matter
i Picsfive/Shutterstock.com; **v** sergeyryzhov/iStock/Getty Images Plus/Getty Images; **vi–vii** MBPROJEKT_Maciej_Bledowski/iStock/Getty Images Plus/Getty Images; **viiib** AlinaMD/iStock/Getty Images Plus/Getty Images; **ix** Wavebreakmedia/iStock/Getty Images Plus/Getty Images; **x** © Globisens; **xi** Cultura Images | Phil Boorman/Media Bakery

Chapter 1
1 alienwormzond/iStock/Getty Images Plus/Getty Images; **4** Khamkhlai Thanet/Shutterstock.com; **5** georgeclerk/iStock/Getty Images Plus/Getty Images; **7** Migrenart/iStock/Getty Images Plus/Getty Images; **9** Caiaimage/Martin Barraud/OJO+/Getty Images; **11** bdspn/Stock/Getty Images Plus/Getty Images; **13** FineArtCraig/iStock/Getty Images Plus/Getty Images; **15** weestock Images/Alamy Stock Photo

Chapter 2
21 Science Photo Library | MARK GARLICK/Media Bakery; **25** © Patricia Hofmeester/123RF; **27** Michal Ludwiczak/Shutterstock.com; **29** Professor25/iStock/Getty Images Plus/Getty Images

Chapter 3
33 © Marilou Trias | Dreamstime.com; **37** Georgijevic/E+/Getty Images; **39** Matthias Ritzmann/Corbis/Getty Images Plus/Getty Images

Chapter 4
41 Viktor_Gladkov/iStock/Getty Images Plus/Getty Images; **45** vchal/iStock/Getty Images Plus/Getty Images; **47** Steve Debenport/iStock/Getty Images Plus/Getty Images

Chapter 5
51 Adobe Stock/Davizro Photography; **55** BIOSPHOTO/Alamy Stock Photo

Chapter 6
59 klosfoto/E+/Getty Images; **61** Turtle Rock Scientific/Science Source; **63** AndreyCherkasov/iStock/Getty Images Plus/Getty Images

Chapter 7
67 chameleonseye/iStock/Getty Images Plus/Getty Images; **70–71bg** alenaohneva/iStock/Getty Images Plus/Getty Images; **73** NatalieIme/iStock/Getty Images Plus/Getty Images

Chapter 8
77 kasezo/iStock/Getty Images Plus/Getty Images; **78** boule13/iStock/Getty Images Plus/Getty Images; **81** New Africa/Shutterstock.com; **85** hh5800/E+/Getty Images; **86, 87** jhorrocks/iStock/Getty Images Plus/Getty Images

Chapter 9
89 TomekD76/iStock/Getty Images Plus/Getty Images; **93** Fox3X/iStock/Getty Images Plus/Getty Images

Chapter 10
97 sneska/iStock/Getty Images Plus/Getty Images; **98** Kannika Suwan/EyeEm/Getty Images; **100** StanislavSalamanov/iStock/Getty Images Plus/Getty Images; **103** LauriPatterson/E+/Getty Images

Chapter 11
107 NicoElNino/iStock/Getty Images Plus/Getty Images; **111** jacobeukman/iStock/Getty Images Plus/Getty Images; **117** Graiki/Moment/Getty Images

Chapter 12
121 technotr/E+/Getty Images; **125** Rodrigo Garrido/Shutterstock.com

Chapter 13
129 Thatree Thitivongvaroon/Moment/Getty Images; **133** IKvyatkovskaya/iStock/Getty Images Plus/Getty Images; **137** Jelena Danilovic/iStock/Getty Images Plus/Getty Images

Chapter 14
139 Australian Scenics/Photolibrary/Getty Images Plus/Getty Images; **143** Kerrick/E+/Getty Images; **145** patpitchaya/iStock/Getty Images Plus/Getty Images

Chapter 15
149 yoh4nn/iStock/Getty Images Plus/Getty Images; **155** Jeremy Hudson/PhotodiscGetty Images

Chapter 16
161 EMPPhotography/iStock/Getty Images Plus/Getty Images; **165** Peter Titmuss/Alamy Stock Photo; **167** Okea/iStock/Getty Images Plus/Getty Images

Chapter 17
169 Henry Westheim Photography/Alamy Stock Photo; **175** Moorefam/iStock/Getty Images Plus/Getty Images; **177** toos/iStock/Getty Images Plus/Getty Images

Chapter 18
179 Pixsooz/Shutterstock.com; **182, 183** lasagnaforone/Digital Vision Vectors/Getty Images; **185** Richard Green/Alamy Stock Photo

Chapter 19
189 Ridofranz/iStock/Getty Images Plus/Getty Images; **190l** Gunter Marx Photography/Corbis NX/Getty Images Plus/Getty Images; **190r** Sarah Lompe; **192** Electronic-Axial-Lead-Resistors-Array by Evan-Amos/Wikimedia Commons/Public Domain; **193** paladaxar/iStock/Getty Images Plus/Getty Images; **195** Chayapon Bootboonneam/EyeEm/Getty Images; **196** Yves Marcoux/First Light/Getty Images Plus/Getty Images; **197** Besiki Kavtaradze/iStock/Getty Images Plus/Getty Images; **198–99** Tevarak/iStock/Getty Images Plus/Getty Images; **201** Corepics VOF/Shutterstock.com; **202** HRAUN/E+/Getty Images; **204** Adobe Stock/makaule

Chapter 20

207 Phil Degginger/Alamy Stock Photo; **210** Ilona Shorokhova/iStock/Getty Images Plus/Getty Images; **211** jatrax/iStock/Getty Images Plus/Getty Images

Chapter 21

213 Tobias Richter/LOOK-foto/Getty Images Plus/Getty Images; **216** EvgeniiAnd/iStock/Getty Images Plus/Getty Images; **218** Diabluses/iStock/Getty Images Plus/Getty Images; **221** M Swiet Productions/Moment Select/Getty Images Plus/Getty Images

Chapter 22

227 sovika/iStock/Getty Images Plus/Getty Images; **230** Takashi Images/Shutterstock.com; **231** photomim/Shutterstock.com; **235** assalve/iStock/Getty Images Plus/Getty Images

Back Matter

237 PhotoAlto Images | Frederic Cirou/Media Bakery; **238**t ognianm/iStock/Getty Images; **238**b ugurhan/E+/Getty Images; **239** Stefan90/iStock/Getty Images; **240–43** Flinn Scientific SDS Sheets/Public Domain; **244** (beaker, flask) choness/iStock/Getty Images; **244** (Bunsen burner) mozcann/E+/Getty Images; **244** (pipette) Paket/iStock/Getty Images; **244** (filter funnel) mozcann/E+/Getty Images; **244** (graduated cylinder) prill/iStock/Gety Images; **244** (hot plate) Martin Shields/Science Source; **244** (iron ring) Rabbitmindphoto/Shutterstock.com; **244** (electronic laboratory balance) Martin Shields/Alamy Stock Photo; **244** (laboratory balance) Science Photo Library/Getty Images; **244** (mortar and pestle) Steven Raniszewski/Getty Images; **245** (pipette) sciencephotos/Alamy Stock Photo; **245** (ring stand) nipastock/iStock/Getty Images; **245** (goggles) Liudmyla Liudmyla/iStock/Getty Images; **245** (spatula) Anak Surasarang/Shutterstock.com; **245** (stirring rod) Paday/iStock/Getty Images; **245** (test tube) ImageDB/iStock/Getty Images; **245** (test tube holder) Rabbitmindphoto/Shutterstock.com; **245** (test tube rack) Nicholas Shkoda/iStock/Getty Images; **245** (wash bottle) Turtle Rock Scientific/Science Source; **245** (watch glass) creativesunday2016/iStock/Getty Images; **245** (wire gauze) Martyn F. Chillmaid/Science Source; **246**t Science Photo Library/Getty Images; **246**b, **247**t Martin Shields/Alamy Stock Photo; **247**br mozcann/E+/Getty Images; **247**bl SPL/Science Source; **248**t RapidEye/iStock/Getty Images; **248**b © NatalieIme/iStock/Getty Images; **251** sergeyryzhov/iStock/Getty Images; **252** (AC charger, USB cable, air pressure tube, start guide, flyer) Hannah Labadorf/© Globisens; **252** (Labdisc data logger, banana cables, temperature probe), **253-61** © Globisens

PERIODIC TABLE OF THE ELEMENTS

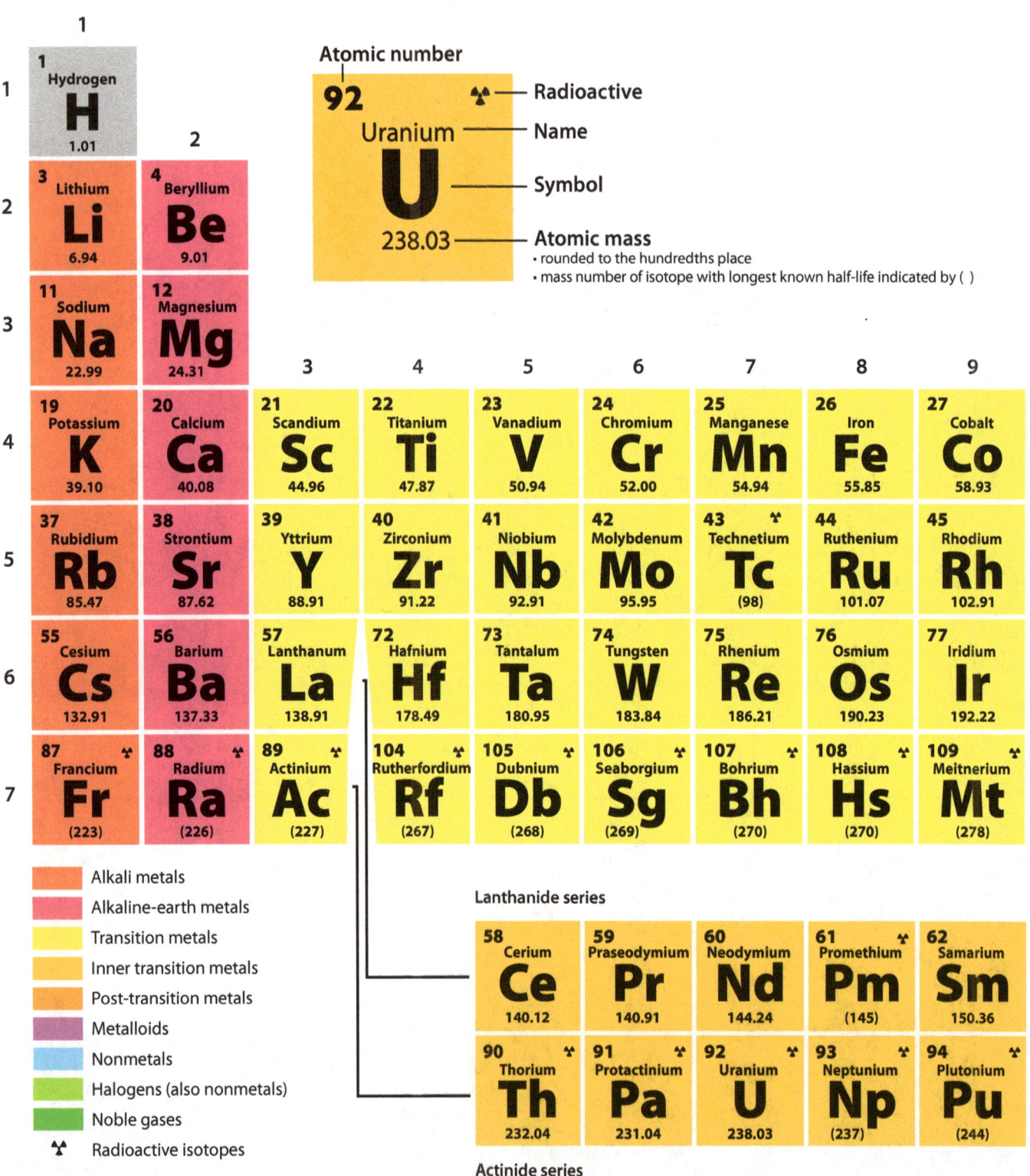

				13	14	15	16	17	18
									2 Helium **He** 4.00
				5 Boron **B** 10.81	6 Carbon **C** 12.01	7 Nitrogen **N** 14.01	8 Oxygen **O** 16.00	9 Fluorine **F** 19.00	10 Neon **Ne** 20.18
				13 Aluminum **Al** 26.98	14 Silicon **Si** 28.09	15 Phosphorus **P** 30.97	16 Sulfur **S** 32.06	17 Chlorine **Cl** 35.45	18 Argon **Ar** 39.95
10	11	12							
28 Nickel **Ni** 58.69	29 Copper **Cu** 63.55	30 Zinc **Zn** 65.38	31 Gallium **Ga** 69.72	32 Germanium **Ge** 72.63	33 Arsenic **As** 74.92	34 Selenium **Se** 78.97	35 Bromine **Br** 79.90	36 Krypton **Kr** 83.80	
46 Palladium **Pd** 106.42	47 Silver **Ag** 107.87	48 Cadmium **Cd** 112.41	49 Indium **In** 114.82	50 Tin **Sn** 118.71	51 Antimony **Sb** 121.76	52 Tellurium **Te** 127.60	53 Iodine **I** 126.90	54 Xenon **Xe** 131.29	
78 Platinum **Pt** 195.08	79 Gold **Au** 196.97	80 Mercury **Hg** 200.59	81 Thallium **Tl** 204.38	82 Lead **Pb** 207.24	83 Bismuth **Bi** 208.98	84 Polonium **Po** (209)	85 Astatine **At** (210)	86 Radon **Rn** (222)	
110 Darmstadtium **Ds** (281)	111 Roentgenium **Rg** (282)	112 Copernicium **Cn** (285)	113 Nihonium **Nh** (286)	114 Flerovium **Fl** (289)	115 Moscovium **Mc** (290)	116 Livermorium **Lv** (293)	117 Tennessine **Ts** (294)	118 Oganesson **Og** (294)	

63 Europium **Eu** 151.96	64 Gadolinium **Gd** 157.25	65 Terbium **Tb** 158.93	66 Dysprosium **Dy** 162.50	67 Holmium **Ho** 164.93	68 Erbium **Er** 167.26	69 Thulium **Tm** 168.93	70 Ytterbium **Yb** 173.05	71 Lutetium **Lu** 174.97
95 Americium **Am** (243)	96 Curium **Cm** (247)	97 Berkelium **Bk** (247)	98 Californium **Cf** (251)	99 Einsteinium **Es** (252)	100 Fermium **Fm** (257)	101 Mendelevium **Md** (258)	102 Nobelium **No** (259)	103 Lawrencium **Lr** (266)

Periodic Table of the Elements